The Complete Irrigation Handbook: Management, Pollution and Solutions

The Complete Irrigation Handbook: Management, Pollution and Solutions

Edited by **Davis Twomey**

New York

Published by Callisto Reference,
106 Park Avenue, Suite 200,
New York, NY 10016, USA
www.callistoreference.com

The Complete Irrigation Handbook: Management, Pollution and Solutions
Edited by Davis Twomey

International Standard Book Number: 978-1-63239-594-8 (Hardback)

Printed in the United States of America.

Contents

Preface

This is an all-inclusive book which provides in-depth information regarding irrigation, its management as well as problems related to pollution along with the solutions. Irrigated agriculture is the leading utilizer of fresh water across the globe and, because of the huge area in use, is one of the primary sources of pollution to the water resources. This book contains comprehensive information which encompasses distinct issues and problems of irrigated agriculture ranging from water use in various irrigated systems to pollution produced by irrigated agriculture. Furthermore, it also presents novel possibilities of latest irrigation methodologies involving reuse of drainage water and wastewater to help in reduction of freshwater extractions. A broad spectrum of issues has been covered in this book, regarding the assessment of irrigated agriculture effects and management practices for the reduction of these impacts on the environment.

The information contained in this book is the result of intensive hard work done by researchers in this field. All due efforts have been made to make this book serve as a complete guiding source for students and researchers. The topics in this book have been comprehensively explained to help readers understand the growing trends in the field.

I would like to thank the entire group of writers who made sincere efforts in this book and my family who supported me in my efforts of working on this book. I take this opportunity to thank all those who have been a guiding force throughout my life.

Editor

Watershed Monitoring for the Assessment of Irrigation Water Use and Irrigation Contamination

Iker García-Garizábal[1,2], Raphael Abrahao[3] and Jesús Causapé[2]
[1]University of Zaragoza,
[2]Spanish Geological Survey
[3]MIRARCO–Mining Innovation, Laurentian University
[1,2]Spain
[3]Canada

1. Introduction

One of the main current questions on the sustainability of life in our planet is if in the next years there will be sufficient water to satisfy the necessities of agriculture and of the other users of this important resource (urban, industrial, touristic and ecological uses).

Irrigation activities allow for the increase of agrarian yields, also allowing for a greater stability in food supply, mainly in those regions where the development of crops is limited by rain. In this way, agriculture consumes 70% of all water extracted from natural courses, being considered the main responsible factor for global fresh water shortage (FAO, 2002).

Nevertheless, although the volumes employed by the agrarian sector are high, at a global level it is estimated that only 50% of the water extracted is finally utilized by plants; the remaining share ends up in drainage and irrigation return flows in rivers and aquifers (FAO, 2003).

These volumes returned to water systems could contribute to a reduction in the impact generated by the extraction of resources if the water quality was not very distant from that of the original water extracted, due to the transport of salts and agrochemicals from the soil profile.

Regarding the presence of agrochemicals, nitrate is a very important issue for water quality, and above all, is associated with notable changes implemented in agriculture in the last decades (OMS, 2004). The problem of nitrate with respect to other agrochemicals is its effect on human health by the simple fact of being present in high concentrations in potable water. The consumption of water with high concentrations of nitrate causes the development of methemoglobinemia in the blood, making the blood stream incapable of transporting enough oxygen through the organism and leading to death of the individual (OMS, 2004).

On the other hand, the occurrence of high concentrations of nitrate in rivers and oceans is causing serious environmental effects on aquatic plants and animals, leading to the occurrence of anoxic zones and eutrophication of water resources (Diaz, 2001), as is evidenced on the coast of the United States (Scavia and Bricker, 2006) and China (Wang, 2006).

The impacts generated by irrigation can be aggravated by physical (geology and climate) and agronomic (management of irrigation and fertilization) factors. For example, the natural salinity of the area in which irrigation is implemented can contribute significantly to the export of salts from the irrigated area, affecting water resources downstream (Christen et al., 2001; Tanji and Kielen, 2002). Strong rain events, on the other side, cause lateral and vertical mobility of the exported masses of contaminants (Thayalakumaran et al., 2007). Intense rain can also contribute to the erosion of soil and leaching of fertilizers and other agrochemicals (Carter, 2000).

Regarding agronomic factors, García-Garizábal et al. (2009) verified that an adequate management of irrigation water can reduce significantly the masses of salts and nitrates exported from an agrarian watershed. Gheysari et al. (2009) indicate that it is possible to control the levels of nitrate leaching from the root zone with an appropriate joint management of irrigation and fertilization. Also, it has been demonstrated that a decrease in nitrogenous fertilization can considerably decrease nitrate leaching levels without causing a drop in productivity (Moreno et al., 1996; Cui et al., 2010). It is therefore possible to achieve equilibrium between acceptable environmental impacts and high agrarian yields.

The main objective of this chapter is to compare and relate water use and contamination generated by salts and nitrates in two irrigated areas with different agronomic characteristics (flood vs. pressurized irrigation). This was carried out through the monitoring of the irrigated hydrological watersheds, analyzing the water use index and salt and nitrate contamination indices calculated for each watershed.

2. Description of study zones

2.1 Location

The study zones correspond to two irrigated watersheds, which are representative of the Bardenas Irrigation District (Spain; Figure 1). The first watershed presents flood irrigation while the second watershed presents pressurized irrigation systems. Both zones are supplied with good quality water (EC = 0.3 dS/m; $NO_3^- = 2$ mg/l) from the Yesa reservoir, transported to the watersheds through the Bardenas channel (Figure 1).

Fig. 1. Location of the Bardenas Irrigation District and the irrigated watersheds, object of this study.

The irrigation ditch network surrounding the flood-irrigated watershed constitutes the superficial water divide, delimiting a 95 ha hydrological watershed of which 96% corresponds to soils destined to irrigation. The remaining surface is occupied by access trails and superficial drainage network, which evacuates the irrigation surplus. The watershed is located at 367 masl

In the case of the pressurized-irrigated area, the watershed was delimited from the terrain digital model (CHE, 2010) and a point situated at the end of the gully, which is a natural drain and evacuates the agrarian drainage waters of the watershed. This watershed presents an extension of 405 ha of irrigated area and is located at an average altitude of 350 masl

2.2 Climate

The climate is Mediterranean warm (ITGE, 1985), presenting a historical reference evapotranspiration (ET_0) of 1068 mm/year and precipitation (P) of 460 mm/year (Figure 2; GA, 2009), with high annual variability. During the three years comprehending this study, there was an average climate year (2006) and two medium-dry years (2007 and 2008).

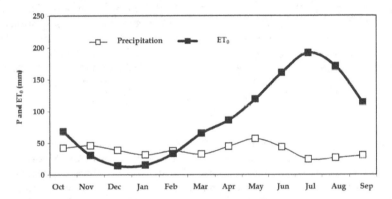

Fig. 2. Historical monthly dynamics of precipitation (P) and reference evapotranspiration (ET_0) in the zone of the two studied watersheds (GA, 2009a).

The dry months correspond to winter and summer seasons, while the wettest months are registered during spring and fall. Regarding ET_0, minimum values are registered in winter and maximum values in summer, which widely exceed precipitation (Figure 2), making irrigation necessary to satisfy the water demands of the crops.

2.3 Geology

The watersheds are located on a glacis of gravel with loamy matrix, constituting a free aquifer. A network of drainage ditches and drains affects the aquifer by forming a valley where tertiary substratum surfaces, constituting the local impermeable limit and acting also as a source of salts (Causapé et al., 2004a).

A sampling network transformed into piezometers determined a gravel thickness of up to 5.5 meters in the flood-irrigation watershed and of up to 10 meters in the pressurized-irrigation watershed. The thicknesses decreased progressively from the topographically higher zones until the lower part of the watershed, where it almost disappeared and the impermeable substratum surfaced.

Regarding the hydraulic characteristics of the aquifer, ITGE (1995) and SIAS (2009) estimate permeabilities of up to 90 m/day and transmissivities of up to 600 m²/day, with an effective porosity of approximately 10-15%.

2.4 Soils

The soils of the study zones were characterized through the elaboration of apparent electric conductivity (ECa) maps in homogeneous humidity conditions close to field capacity (after intense rain). To this end, ECa readings were obtained with a georeferenced mobile electromagnetic sensor (SEMG; Amezketa, 2007) model IS of Dualem, in horizontal configuration (ECah), integrating a depth of one meter, and in vertical configuration (ECav), which integrates a depth of 2 meters.

Data revealed low soil salinity (CEah$_{Flood}$= 0.16 dS/m; CEav$_{Flood}$= 0.25 dS/m; CEah$_{Press}$= 0.27 dS/m; CEav$_{Press}$= 0.48 dS/m), although slightly higher values were found in the soil of the pressurized-irrigation zone due to the natural salinity of the subsoil. The highest ECa recorded in the flood-irrigation watershed was 1.28 dS/m on tertiary lutites compared to almost 6 dS/m (Urdanoz et al., 2008) registered on the tertiary of the pressurized-irrigation zone.

Regarding the texture of the soils, Lecina et al. (2005) have already made a first characterization in the zone, classifying the soils in two groups. The first group corresponds to the soil developed on the glacis, with loamy texture, stone content between 11 and 13% and a moderate water holding capacity (WHC), classified as Calcixerollic Xerochrept with Petrocalcic Xerochrept inclusions (Soil Survey Staff, 1992). Conversely, the second group included the soil developed on the tertiary with loamy texture, much lower stone content, between 4 and 18%, and a higher water holding capacity, and classified as Typic Xerofluvent (Soil Survey Staff, 1992).

2.5 Agronomy: Irrigation and fertilization

Irrigation is the main component differentiating the two watersheds (Figure 3). Therefore, although both watersheds present on-demand irrigation, in which the farmers chose the time and amount of water to be applied (maximum annual water allowances established at the beginning of the season in function of the available reserves in the reservoir supplying the system), one of the watersheds was submitted to flood irrigation while the other watershed presented pressurized irrigation systems, with 86% of the surface occupied by sprinkler systems and the remaining 14% occupied by drip irrigation systems.

Fig. 3. Pictures of the flood irrigation system (A) and pressurized irrigation system (B).

Regarding the crops, distribution varied significantly in the watersheds as a consequence of the irrigation system in use. In the flood-irrigation watershed, winter cereal (46%) and alfalfa (31%) were the main crops, at the expense of minority crops such as maize and sunflower, with extensions no greater than 15% of the annual surface (Table 1). In the pressurized-irrigation watershed, maize was always the major crop (55%), followed by winter cereal (24%) and tomatoes (9%), with minor contributions of broccoli, sunflower or peas. Alfalfa was not found in the second watershed, even though it is a very common crop in this Irrigation District.

%	2006		2007		2008	
Crop	Flood	Pressurized	Flood	Pressurized	Flood	Pressurized
Winter cereal	33	--	51	25	55	23
Alfalfa	39	--	31	--	24	--
Maize	8	61	3	63	--	40
Tomato	--	10	--	4	--	14
Others	20	29	15	8	21	23

Table 1. Distribution of the main crops in the flood-irrigation watershed and in the pressurized-irrigation watershed, for the three hydrological study years (2006-2008).

Irrigation volumes present variations among crops. In the flood-irrigation watershed, winter cereal presented 2-3 irrigation doses per year, each one of 128 mm. It must be noted that, very punctually, some farmers did not irrigate because rain was sufficient to satisfy the water demands. Alfalfa and maize, with higher water demands, presented 8-10 irrigation doses of 122 mm and 8 irrigation doses of 136 mm, respectively.

	Flood			Pressurized		
Crop	Irrigation	N Fert.	Yield	Irrigation	N Fert.	Yield
	mm	kg N/ha	kg/ha	mm	kg N/ha	kg/ha
Winter cereal	235	162	5000	157	164	4600
Alfalfa	1057	61	11700	--	--	--
Maize	1088	420	10600	740	380	12000
Tomato	--	--	--	552	182	80000

Table 2. Irrigation doses, nitrogenous fertilization and average yield for the crops in the flood- and pressurized-irrigation watersheds, for the three hydrological study years (2006-2008).

In the pressurized system, irrigation was characterized by a high number of applications, but with small volumes. Therefore, low doses were applied to winter cereal (10 doses of 15.7mm) while corn (40 doses of 18.5 mm) reached a total volume of 740 mm per year. Both were irrigated with sprinkler systems. Tomatoes were irrigated via drip irrigation, with very frequent applications of small doses of water throughout the entire cycle, resulting in a total annual volume of 552 mm.

Regarding fertilization, the average annual doses were 156 kg N/ha in the flood-irrigation watershed and 273 kg N/ha in the pressurized-irrigation watershed, without significant variations in doses for the same crop. Therefore, the doses were sensibly high for corn, 420 kg N/ha in flood systems, compared to 380 kg N/ha in sprinkler systems. Winter cereal received an average fertilization of 163 kg N/ha, while tomatoes received 182 kg N/ha. For alfalfa, the average annual doses of nitrogen reached 61 kg N/ha although this fertilizer was not needed because alfalfa is a leguminous. In this sense, the good agrarian practice code (BOE 1996; BOA 1997), derived from the European directive 91/676 (EU 1991), establishes that nitrogenous fertilization of alfalfa is null with an exception for the year of implementation of the crop, with a limit of 30 kg N/ha. Nitrogenous fertilization was applied mainly in the form of complex NPK fertilizers (8-15-15 and 15-15-15), urea (46% N), nitrogenous solution N-32 (32% N) and, to a smaller extent, ammonia nitrate (33.5% N).

3. Methodology

Water use management was evaluated along with the contamination generated by both irrigated zones during three hydrological years (2006-2008). To this end, annual water balances were executed and the contaminant exports (masses of salts and nitrates) were quantified in each watershed. Subsequently, a series of indices was calculated to evaluate irrigation management and relate the contaminants to the salinity characteristics and nitrogenous fertilization (agronomic) of each irrigated zone. The Irrigation Land Environmental Evaluation Tool (in Spanish EMR; Causape, 2009) was used, which automates the calculations for the execution of the water balances and calculations of water use management indices (net hydric needs-HN; water use index-WUI; irrigation efficiency IE) and contamination indices (salt contamination index-SCI; nitrate contamination index-NCI).

3.1 Water balances

Annual water balances were executed from measurements or estimations of the main inputs, outputs and water storage in each irrigated watershed (Figure 4). The equation used in the balances was:

$$\text{Inputs} - \text{Outputs} - \text{Storage} = \text{Error balance}$$

$$(P + I + IWF) - (ET + Q + EWDL) - (Ss + Sa) = \text{Error} \tag{1}$$

were the inputs through precipitation (P), irrigation (I) and incoming water flows (IWF), minus the outputs through evapotranspiration (ET), drainage (Q) and losses due to evaporation and wind drift and evaporation losses from sprinkler irrigation (EWDL), minus water storage in the soil (Ss) and aquifers (Sa), constitute the balance error.

Climate data regarding precipitation and reference evapotranspiration (ET_0; Penman-Monteith) necessary for the execution of balances were obtained from agro-climatic stations that the Integral Counseling Service to Irrigation (in Spanish SIAR; GA, 2009a) installed in the proximity of the watersheds.

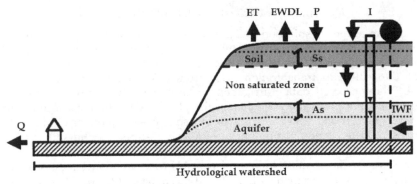

Fig. 4. Hydrogeological conceptual model in which the main hydric components are represented in the irrigated watersheds: irrigation (I), precipitation (P), losses due to evaporation and wind drift of sprinkler irrigation systems (EWDL), evapotranspiration (ET), soil drainage (D), incoming water flows (IWF), flow measured at the gauging station (Q), water stored in the soil (Ss) and in aquifers (Sa).

The daily irrigation volumes were facilitated by the Irrigation district. In the case of pressurized-irrigation, the losses via evaporation and wind drift in sprinkler systems were quantified through the equation proposed by Playán et al. (2005):

$$EWDL\ (\%) = 20.34 + 0.214 \cdot ws\ [m/s]^2 - 2.29 \cdot 10^{-3} \cdot HR\ [\%]^2 \tag{2}$$

where data on wind speed 2 m above the surface (WS, m/s) and relative humidity 1.5 m above ground level (HR, %) were needed.

The annual contribution of incoming water flows to the balance in the flood-irrigation watershed was quantified through the piezometer network. To this end, the saturated water thickness (SWT) measured once every 21 days in a piezometer installed northwest of the watershed was related to the water volume flowing through the drainage at the gauging station. The gauging station presented a rectangular flow meter and electronic limnigraph that registered water height every 15 minutes (h), transformed into flow according the equation (Q_{Flood} (m³/s) = 0.0002 h² – 0.0020 h – 0.0179; n= 9; R^2 = 0.99; p< 0.001), yielding the calculation of IWF_{Flood}:

$$IWF_{Flood}\ (m^3/day) = 186.39\ exp^{1.82\,SWT};\ n = 11;\ R^2 = 0.79;\ p< 0.001 \tag{3}$$

The incoming water flows in the pressurized-irrigation watershed from the unirrigable area included in the watershed were estimated from precipitation data and based on a runoff coefficient of 0.087. This coefficient was obtained from the relationship precipitation-flow. Based on the entire dataset available from the gauging station, it was verified that heavy rains yielded a higher runoff coefficient (0.313), which was then applied to daily rainfall events exceeding 25 mm.

In the pressurized-irrigation watershed, the equation utilized was provided by software Winflume (Wahl, 2000):

$$Q_{Press}\ (m^3/s) = 1.73 \cdot (h + 0.00347)^{1.624}\ for\ h \le 0.5\ m \tag{4}$$

$$Q_{Press} \ (m^3/s) = 10.28 \cdot (h + 0.01125)^{1.725} \text{ for } h > 0.5 \ m \tag{5}$$

Regarding crop evapotranspiration (ET_C), it was calculated on a daily basis from the crop coefficients (K_C) determined for the study zone by Martínez-Cob (2004) and by ET_0 according to the equation $ET_C = ET_0 \cdot K_C$ (Allen et al., 1998). In this sense, ET_C was corrected daily by the real evapotranspiration (ET_R) from calculations developed by the EMR software. Therefore, daily data of irrigation, precipitation, evapotranspiration, along with hypothetic initial useful water available for the plants (AWinitial), constituted the inputs for the execution of the water balance in the soil of each plot, resulting in the daily estimations of real evapotranspiration, useful water stored in the soil (AU) and soil drainage.

Therefore, starting from an initial volume of available water for plants in the soil (AW), EMR adds the daily inputs by irrigation (I - EWDL) and precipitation (P), and ET_C is subtracted only if there is sufficient AW in the soil. EMR considers that ETa = ET_C if AWinitial + P + I - EWDL > ET_C, but otherwise ETa= AWinitial + P + I - EWDL – hence, the soil has a wilting point level of humidity at the end of each day (AW = 0). On the other hand, if AWinitial + P + I - EWDL - ETa > WHC, the program interprets that the field soil capacity has been surpassed, obtaining drainage (DSWB) equal to DSWB = AWinitial + P + I - EWDL - ETa - WHC, leaving the soil at the termination of each day at field capacity (maximum AW = WHC).

In order to obtain an approximate value of the water content in the soil in the beginning of the study, the execution of balances started one year before. With the information generated by the soil water balance, EMR estimates the direct components of the water balance in the watershed: real evapotranspiration, water storage and soil drainage. The drainage volume proceeding from irrigation (DI) was estimated by considering for the days and plots with drainage that if AW + P – ETa ≥ WHC then DI = I - EWDL and otherwise DI = [I - EWDL] - [WHC - (AW + P - ETa)]. The interpretation of this calculation is that, on any given day, rainfall will always occur before irrigation and thereby irrigation drainage takes priority over rainfall drainage. It is assumed in this study that a farmer takes rainfall into account when deciding whether to irrigate, although evidently weather forecasting is by no means infallible.

Regarding water storage, from the balance equation it was obtained that soil storage resulted from the difference between water volume at the beginning and end of each hydrological year for each balance estimated by EMR. For water storage in the aquifer, this was calculated from the water height variation in the aquifer, measured by the piezometer network at the beginning and end of each hydrological year, applying an effective porosity between 15-20% according to the lithology of the materials extracted during sampling and to values registered during other local studies (Custodio & Llamas, 1983; ITGE, 1995).

Finally, the adequate closure of the water balances was quantified through the calculation of percentage errors:

$$\text{Error } (\%) = [(\text{Inputs} - \text{Outputs} - \text{Storage}) \ / \ (\text{Inputs} + \text{Outputs} + \text{Storage})] \cdot 200 \tag{6}$$

3.2 Evaluation of water use and irrigation quality

In order to calculate the irrigation quality during the three study years (2006-2008), the net hydric needs (HN) of the crops were calculated along with water use and irrigation efficiency indices, calculated by EMR once acceptable and satisfactory errors were achieved, which highlight the goodness of the water balances.

The hydric needs estimates the volume of irrigation water necessary to avoid crops from suffering water stress and for the soil to contain the same initial moisture conditions. The potential evapotranspiration and final useful water in the soil are added, to which effective precipitation and initial useful water in the soil are subtracted.

$$HN \text{ (mm)} = (ET_C + AW\text{final}) - (AW\text{initial} + Pef) \tag{7}$$

The water use index quantifies the percentage of water resources (irrigation and precipitation) that have been used for evapotranspiration:

$$WUI \text{ (\%)} = [1 - (D + EWDL) / (I + P)] \cdot 100 \tag{8}$$

Finally, irrigation efficiency evaluates the percentage of irrigation volume that has not left the system, being used to satisfy the hydric needs of the crops or stored in the water storage in the soil.

$$IE \text{ (\%)} = [1 - (DI + EWDL) / (I)] \cdot 100 \tag{9}$$

3.3 Irrigation contamination: Masses of salts and nitrates exported

In order to quantify the masses of contaminants exported through the drainage associated with the watershed, salt and nitrate concentrations were assigned to the superficial drainage, to the subterranean flow, and to water storage in the aquifer.

$$D = Q - IWF + Sa \tag{10}$$

To this end, drainage stations were equipped with automatic water sampling equipment, programmed to collect daily samples. Subsequently the water samples were taken to the laboratory where the electrical conductivity at 25 °C was determined with an Orion 5-star conductivimeter equipped with a DuraProbe probe, and nitrate concentration was determined via colorimetry (AutoAnalyzer 3).

In order to determine the salt concentration in each water sample, electrical conductivity was transformed into total dissolved solids with the equation:

$$TDS \text{ (mg/l)} = DR \text{ (mg/l)} + \tfrac{1}{2} HCO_3^- \text{ (mg/l); (Custodio y Llamas, 1983)} \tag{11}$$

being DR the dry residue, and HCO_3^- the concentration of bicarbonate measured in 31 and 17 samples of the flood- and pressurized-irrigation watersheds, respectively. The values calculated were related to the measured EC for each water sample analyzed and were used to calculate the total dissolved solids in the drainage waters of both irrigated areas:

$$TDS_{Flood} \text{ (mg/l)} = 704 \cdot EC \text{ (dS/m)} + 90; n = 31; R^2 = 0.97; p < 0.001 \tag{12}$$

$$TDS_{Press} \text{ (mg/l)} = 712 \cdot EC \text{ (dS/m)} - 105; n = 17; R^2 = 0.99; p < 0.001 \tag{13}$$

For the incoming water flows, the electrical conductivity and nitrate concentration were determined from monthly values of water samples collected at subterranean or superficial entry points. Finally, the masses of salts and nitrates stored in the aquifer were obtained from the analyses of manually-obtained samples taken October 1 of the corresponding year for each piezometer.

3.4 Salt and nitrate contamination indices

Habitually, water contamination is evaluated by the contaminant concentration, although it is the load of exported salts in irrigation return flows that modifies the salinity of the hydric systems receiving such return flows, in function of the mixture proportions. Nevertheless, when considering only the masses of salts exported, "natural" salinity can mask the salinity induced by the management of each irrigation zone. The salt contamination index was calculated (SCI; Causapé, 2009) to determine the environmental impact of irrigation, and compare it to other zones with different natural conditions. This index corrects the exported mass by the electrical conductivity of drainage water under nonirrigated conditions (EC_{NR}), which is an indicator that represents the "natural" salinity of each irrigated zone.

$$SCI = D_{Salts} / EC_{NR} \qquad (14)$$

In the case of nitrate, the exported mass is conditioned by the crops, hindering the comparison of the agroenvironmental impact induced by different irrigated zones or different years of the same irrigated area. The nitrate contamination index (NCI; Causapé, 2009) allows for such comparisons, differentiating the crop pattern with respect to other variables such as climate or agronomic management (irrigation and fertilization). This index analyzes the impact of agrarian activities and fertilization practices through a relationship between the nitrate exported through the drainage of the watershed and the theoretical nitrogenous fertilization needs (FN = Average yield (GA, 2009b) · Nitrogen extractions (Orús and Sin, 2009)) of the area to evaluate.

$$NCI = D_N / FN \qquad (15)$$

4. Results and discussion

4.1 Water balances

Water balances resulted satisfactory due to annual errors between -4.4% and 0.3% (Table 3), which remark the goodness of the balances and an adequate measurement and/or estimation of the components. In this way, it was possible to carry out the calculation of the management indices from the different components that constitute the balance equation.

Irrigation constituted the main contribution of water to the watersheds (45% of inputs), except in 2006 and 2007 for the pressurized-irrigation zone, where the installation of the irrigation systems was still being carried out and a part of the irrigable plots was under fallow conditions and did not present water supply. In 2008, after total implementation of irrigation systems, the water doses applied by the farmers increased until the same magnitude order was achieved for both watersheds (Table 3).

Regarding precipitation, it is considered to be the second most important water input in the balances (41%), oscillating between 426-450 mm in the rainiest year (2006) and 305-361 mm in the driest year (2008). Finally, subterranean water flows constituted up to 24% of the water inputs involved in the balances.

Regarding the outputs, evapotranspiration was the main component, resulting in 63-78% of outputs. The water volume measured in the drainage stations varied significantly, constituting 37% of outputs in the watershed with flood irrigation, and 16% in the watershed irrigated by pressurized systems. Nevertheless, subterranean flows presented a greater contribution in the flood-irrigation watershed (Table 3). When discounting the contributions of subterranean water to the volume of water flowing through the gauging station, the drainage of the watershed was always greater in the flood-irrigation system. The water outputs counted as evaporation and drift losses increased to 6%.

	Inputs			Outputs			Storage			
	P	I	IWF	ET	Q	EWDL	Sa	Ss	Unb.	Error
	mm	mm	mm	mm	mm	mm	mm	mm	mm	%
Flood										
2006	450	567	285	830	417	0	42	65	-52	-4.4
2007	372	512	307	753	469	0	4	-39	4	0.3
2008	305	559	271	686	451	0	-16	13	1	0.1
Total	1127	1638	862	2269	1337	0	30	39	-48	-1.5
Pressurized										
2006	426	144	56	425	123	20	9	48	1	0.2
2007	411	397	31	643	106	57	68	-36	1	0.1
2008	361	519	27	656	118	59	69	7	-2	-0.2
Total	1198	1060	114	1724	347	136	146	19	0	0.0

Table 3. Water balance in the flood- and pressurized-irrigation watersheds. Inputs [precipitation (P), irrigation (I), incoming water flows (IWF)], outputs [evapotranspiration (ET), gauging station (Q) and evaporation and wind drift losses in sprinkler irrigation systems (EWDL)] and storage [in soil (Ss) and aquifers (Sa)] of water during the study period 2006-2008). Balance error (inputs-outputs-storage and unbalance).

Water storage was only 5-10% of the water volume involved in the balances, although its consideration resulted important in this type of studies as in the case of soil, WHC (maximum water volume that can be stored/evacuated from soil) itself can be in the order of precipitations during the driest years. The flood-irrigation watershed presented small annual variations in aquifer storage, while the pressurized-irrigation watershed presented water storage in the aquifer for all years, possibly associated with the fact that the pressurized watershed had newly-implement systems and did not reach equilibrium conditions at phreatic levels. In this sense, it is predicted that in the next years the variations in storage will decrease until equilibrium conditions are achieved in the system, and in the future both zones will probably present similar storage variations.

4.2 Evaluation of water use and irrigation quality

Evapotranspiration evolved differently in the two watersheds during the study period. In the flood-irrigation watershed, evapotranspiration suffered a decrease of 17% due to a change in crop pattern, with an expansion of winter cereal at the expense of maize and alfalfa. In the pressurized-irrigation watershed, evapotranspiration increased by 54%, due to the progressive increase of cultivated surface once the installation of irrigations systems was completed. Crop variations in the watersheds are reflected also on the hydric needs of the system, although unit volumes were similar, the greater cultivated surface in the pressurized-irrigation watershed conditioned higher water demands (Table 4).

| Flood/Pressurized | HN | I | IE | | | | WUI |
			Winter cereal	Maize	Alfalfa	Tomatoe	
Year	hm³/year	hm³/year	%	%	%	%	%
2006	0.54 / 0.54	0.55 / 0.58	82 / --	74 / 75	78 / --	-- / 90	87 / 85
2007	0.40 / 1.33	0.48 / 1.61	88 / 62	56 / 74	77 / --	-- / 90	82 / 84
2008	0.55/ 1.90	0.53/ 2.10	82 / 86	-- / 71	72 / --	-- / 86	79 / 83
Average	0.50 / 1.26	0.52 / 1.43	84 / 74	65 / 73	76 / --	-- / 89	83 / 84

Table 4. Hydric needs (HN) of the crops, irrigation volume (I), irrigation efficiency (IE) of the main crops, and water use index (WUI) in the two irrigated watersheds during hydrological years 2006-2008.

The water use index was moderate-high, reaching 83% in the flood-irrigation watershed and 84% in the pressurized-irrigation watershed (90% could have been reached if evaporation and wind drift losses in the sprinkler system were nil). Evaporation and wind drift losses accounted for 13% of total irrigation in the watershed (15% of sprinkler irrigation). This value is slightly inferior to that calculated by Dechmi et al. (2003) and Playán et al. (2005) in other sprinkler-irrigation zones in the proximities, where evaporation and wind drift losses accounted for 15-20% of the applied irrigation.

Tomatoes presented the best irrigation applications, achieving plot irrigation efficiencies of 89%, followed by winter cereal (79%) and alfalfa (76%). Maize, which presents a high economic value in this zone, presented the lowest efficiency values (69%), possibly due to the fact that the great volumes of water were applied by the farmers when faced by the possibility of low productivity due to hydric deficit.

In this sense, although the higher efficiency and better use of water is demonstrated in pressurized-irrigation systems under adequate agronomic management (Clemmens & Dedrick, 1994; Zalidis et al., 1997; Tedeschi et al., 2001; Al-Jamal et al., 2001; Caballero et al., 2001; Cavero et al., 2003; Causapé et al., 2006;) in comparison to nonpressurized- or flood-irrigation systems (Clemmens & Dedrick, 1994; Isidoro et al., 2004; Causapé et al., 2004b; Causapé et al., 2006), an adequate flood irrigation management has allowed for water resource use values similar to those of an adequately managed modern pressurized system (Table 4). García-Garizábal & Causapé (2010) verified that the implementation of simple improvements in flood irrigation management on the part of the irrigation management organisms (from rotation to on-demand flood irrigation with maximum water allowances, and creation of water consumption accounts) increased by 26% the water use at the Irrigation District (Table 5).

	Water flow	Irrigation efficiency
Year	hm³	%
2000	133	67
2007	116	93

Table 5. Water volume circulating through the drainage network of Bardenas District n° V and irrigation efficiency in 2000 (traditional flood irrigation) and 2007 (improved flood irrigation). Taken from García-Garizábal & Causapé (2010).

Therefore, an adequate management of flood irrigation and the implementation of pressurized systems allowed for good water use indices to be obtained by the farmers (79-87%), although water management still has to be sequentially adjusted to achieve continuous and uniform high values at the plots, which could be up to 95% (Tanji & Kielen, 2002). Superior efficiency recordings are not recommended, as the good conservation state of the agrarian soils would be at risk due to the insufficient leaching of evapoconcentrated salts accumulated in the soil profile (Abrol et al., 1988).

Therefore, in accordance to the previous results, the water use indices in the flood-irrigation watershed have a scarce margin for improvement, and the farmers' labour should be focused on maintaining such indices. Nevertheless, Lecina et al. (2005) affirm that it could be possible to increase water use in this zone with the implementation of pressurized-irrigation systems. On the other hand, the farmers of the watershed that already presents pressurized irrigation systems must concentrate efforts on improving irrigation application, mainly reducing the losses through evaporation and wind drift in sprinkler systems by applying water during the night or in low-wind periods (Playán et al., 2005; Zapata et al., 2007; Zapata et al., 2009).

4.3 Irrigation contamination: Exported masses and contamination indices

The salt masses accounted at the gauging stations of both watersheds were similar, although the pressurized-irrigation watershed presented a greater annual variability due to the increase in drainage volumes (irrigation system under implementation) and to the higher salinity of its return flows. The salts exported by each watershed (own drainage of the system) were significantly different, with 1.7 t/year in the flood-irrigation watershed and 3.2 t/year in the pressurized-irrigation watershed, due to less salt masses incorporated in subterranean flows and higher storage of salts in the aquifer in the latter (Table 6).

In relation to other zones, the masses exported by both watersheds were lower than those measured in irrigation zones with low-moderate water use index (around 50%), presenting annual exports between 3.4 and 4.7 t/year (Causapé et al., 2004c; Duncan et al., 2008), and similar to values encountered in irrigation zones with moderate-high irrigation efficiencies, between 73% (5.2 t/ha year; Roman et al., 1999) and 82% (3.9 t/ha year; Caballero et al., 2001).

Flood / Pressurized	Q_{Salts}	IWF_{Salts}	Sa_{Salts}	D_{Salts}	CE_{NR}	SCI
Year	t/ha	t/ha	t/ha	t/ha	dS/m	$t \cdot ha^{-1}/dS \cdot m^{-1}$
2006	2.8 / 2.8	1.4 / 1.8	0.5 / 1.2	1.9 / 2.2	1.1 / 3.8	1.8 / 0.6
2007	3.1 / 2.3	1.6 / 0.8	0.0 / 2.0	1.5 / 3.5	1.1 / 3.8	1.4 / 0.9
2008	2.9 / 3.2	1.2 / 0.6	-0.1 / 1.3	1.6 / 3.9	1.1 / 3,8	1,5 / 1,0
Average	2.9 / 2.8	1.4 / 1.1	0.1 / 1.5	1.7 / 3.2	1,1 / 3.8	1.6 / 0.8

Table 6. Mass of salts exported through drainage (Q_{salts}), mass of salts introduced in the incoming subterranean water flows (IWF_{Salts}), mass of salts stored in the aquifer (Sa_{salts}), mass of salts associated with the watershed (D_{salts}),electrical conductivity under nonirrigated conditions (EC_{NR}) and salt contamination index (SCI) in the two studied watersheds (Flood- and pressurized-irrigation).

Regarding the salt contamination index, although the flood-irrigation watershed presented lower natural salinity ($EC_{NR-Flood}$= 1.1 dS/m vs. $EC_{NR-Press}$= 3.8 dS/m) the SCI values were higher than those calculated for the pressurized-irrigation watershed. The high natural salinity of the pressurized-irrigation watershed motivated higher salt exports than those of the flood-irrigation zone, even presenting similar water use values. The higher amount of salts present in the subsoil of the pressurized-irrigation watershed caused the "evaluation" of the salinity impact to be lower due to the impossibility to export naturally low salt masses. In this sense, irrigation zones with high use values obtain salt contamination indices of only 0.4 t/ha year dS/m, while irrigation zones with lower efficiencies present higher values (1.9 t/ha year dS/m), reaching up to 11.4 t/ha year dS/m in agrarian systems with high natural salinity values.

In the case of nitrate, the mass exported by the flood-irrigation zone reached 61 kg N/ha year with a low annual variability, compared to 12 kg N/ha year of the pressurized-irrigation system, although the latter increased exports in more than 200% during the study period (2006-2008).

This increase in the mass of exported nitrates is associated with the increase in drainage volumes measured in the gauging station, due to the expansion of irrigation in the pressurized-irrigation watershed, and to the consequent higher volumes of irrigation water and nitrogenous fertilization entering the watershed.

The flood-irrigation watershed presented minimum annual variations in the nitrate stored in the aquifer, while the pressurized-irrigation watershed always recorded positive nitrate storage due to water storage suffered by the aquifer at the end of irrigation cycles (Table 7).

When compared to other irrigated zones, the masses of exported nitrates were always lower than those quantified in irrigation areas with efficiencies of approximately 50% (Causapé et al., 2004b; Isidoro et al., 2006b) and were similar to the masses measured in irrigation systems with efficiencies higher than 70% (Cavero et al., 2003; Bustos et al., 2006). Nevertheless, in the last two irrigation zones the fertilization needs were of the same order of those calculated in the pressurized-irrigation watershed and 2-3 times superior to those of the flood-irrigation watershed.

Flood / Pressurized	Q_N	IWF_N	Sa_N	D_N	FN	NCI
Years	kg N/ha	kg N/ha	kg N/ha	kg N/ha	kg N/ha	--
2006	59 / 6	10 / 0.4	4 / 4	53 / 10	75 / 78	0.70 / 0.12
2007	67 / 10	10 / 0.6	2 / 26	59 / 36	82 / 150	0.72 / 0.24
2008	56 / 19	6 / 0.4	-3 / 15	47 / 33	77 / 166	0.61 / 0.20
Average	61 / 12	9 / 0.5	1 / 15	53 / 26	78 / 131	0.68 / 0.20

Table 7. Nitrate masses exported through drainage (Q_N), nitrate mass introduced in the incoming subterranean flows (IWF_N), nitrate mass stored in the aquifer (Sa_N), mass of exported nitrates associated with the watershed (D_N) nitrogenous fertilization needs (FN), nitrate contamination index (NCI) in the two studied watersheds (Flood- and pressurized-irrigation).

The lower nitrogenous fertilization needs of the flood-irrigation watershed (NF_{Flood}= 78 kg N/ha vs. NF_{Press}= 131 kg N/ha) induced NCI values always higher than those of the pressurized-irrigation watershed. Therefore, although the flood-irrigation watershed presented lower nitrogen requirements, the masses of exported nitrates was higher due to greater drainage volumes, even when water use indices were similar for both watersheds.

This fact caused a lower impact, although the amount of nitrogen applied to the pressurized-irrigation watershed was higher. This behaviour is similar to the one obtained when nitrate contamination indices are compared to those of other irrigation areas.

Irrigation areas with high water use (73-90%) and nitrogenous fertilization needs (144-213 kg N/ha) obtain nitrate contamination indices of approximately 0.25, while other zones with application efficiencies around 50% present higher NCI values (1.2), presenting in this case nitrogenous fertilization needs of 164 kg N/ha.

5. Conclusions

The proposed methodology for the monitoring of hydrological watersheds and execution of water balances to evaluate irrigation management resulted satisfactory, mainly when calculating annual errors between -4.4% and 0.3%, which remarks the goodness of the balances and allows for the evaluation of irrigation and water resource management from the values provided.

Although there was a clear difference between the irrigation systems present in the evaluated watersheds (flood and sprinkler irrigation), the water use values obtained were similar, approximately 84%. This fact highlights the possibility of reaching adequate water management levels by adapting the irrigation systems, although it is necessary to know the soil, crop, and supply capacity characteristics in order to establish the management strategy and most adequate water management.

Regarding the exports of contaminants, the highest mass of salts was measured in the irrigation zone with the most saline subsoil ($D_{Salts-Flood}$= 1.7 t/ha year $vs.$ $D_{Salts-Press}$= 3.2 t/ha year), while the highest nitrate mass was measured at the watershed with the lowest nitrogenous fertilization input ($D_{Nitrate-Flood}$= 53 kg N/ha year $vs.$ $D_{Nitrate-Press}$= 26 kg N/ha year) due to the greater drainage volume ($Station_{Flood}$= 446 mm $vs.$ $Station_{Press}$= 116 mm). The contamination indices always resulted better for the pressurized irrigation watershed (SCI= 0.8 t · ha^{-1}/dS · m^{-1}; NCI= 0.20) than for the flood-irrigation watershed (SCI= 1.6 t · ha^{-1}/dS · m^{-1}; NCI= 0.68), and therefore it is possible to reduce the degree of contamination if water use is improved, decreasing the irrigation return flow volumes.

6. Acknowledgments

The authors are thankful to the support of the Spanish Government through projects AGL2005-07161-C05-01 (MEC), CGL2009-13410-C02-01 (MICINN) and formation scholarship BES-2006-12662 (MEC) and to the support of European Union Program of High Level Scholarships for Latin America through Alβan Scholarship n°. E07D400318BR. Thanks are extended to the Bardenas Canal Irrigation District n° V Authority, Bardenas Canal Irrigation District n° XI Authority, Agrifood Research and Technology Centre of Aragón (CITA-DGA) and to the farmers for their valuable collaboration.

7. References

Abrol, I.P.; Yadav., J.S.P. & Massud, M.I. (1988). *Salt affected soils and their management*. Food and Agriculture Organization of the United Nations, ISBN 92-5-102686-6, Roma, Italy.

Al-Jamal, M.S.; Ball, S. & Sammis, T.W. (2001). Comparison of sprinkler, trickle and furrow irrigation efficiencies for onion production. *Agricultural Water Management*, Vol.46, No.3, (January 2001), 253-266, ISSN 0378-3774.

Allen, R.; Pereira, L.; Raes, D. & Smith, M. (1998). *Crop evapotranspiration. Guidelines for computing crop water requirements*. Food and Agriculture Organization of the United Nations, ISBN 92-5-104219-5, Roma, Italy.

Amezketa, E. (2007). Soil salinity assessment using directed soil sampling from a geophysical survey with electromagnetic technology: a case study. *Spanish Journal of Agricultural Research*, Vol.5, No.1, (March 2007), 91-101, ISSN 1695-971X.

BOA (1997). *Decreto 77/1997, de 27 de mayo, del Gobierno de Aragón, por el que se aprueba el Código de Buenas Prácticas Agrarias de la Comunidad Autónoma de Aragón y se designan determinadas áreas Zonas Vulnerables a la contaminación de las aguas por los nitratos procedentes de fuentes agrarias. Boletín Oficial de Aragón del 11 de junio de 1997*, Government of Aragon, Spain.

BOE (1996). *Real Decreto 261/1996, de 16 de febrero, sobre protección de las aguas contra la contaminación producida por los nitratos procedentes de fuentes agrarias. Boletín Oficial del Estado del 11 de marzo de 1996*, Government of Spain, Spain.

Caballero, R.; Bustos, A. & Roman, R. (2001). Soil salinity under traditional and improved irrigation schedules in central Spain. *Soil Science Society of America Journal*, Vol.65, No.4, (July-August 2001), 1210-1218, ISSN 0361-5995.

Carter, A.D. (2000). Herbicide movement in soils: principles, pathways and processes. *Weed Research*, Vol.40 (February 2000), 113-122, ISSN 1365-3180.

Causapé, J. (2009). A computer-based program for the assessment of water-induced contamination in irrigated lands. *Environmental Monitoring and Assessment*, Vol.158, No.1-4, (November 2009), 307-314, ISSN 0167-6369.

Causapé, J.; Auqué, L.; Gimeno, M.J.; Mandado, J.; Quílez, D. & Aragüés, R. (2004a). Irrigation effects on the salinity of the Arba and Riguel rivers (Spain): Present diagnosis and expected evolution using geochemical models. *Environmental Geology*, Vol.45, No.5, (March 2004), 703-715, ISSN 0943-0105.

Causapé, J.; Quílez, D. & Aragüés, R. (2004b). Assessment of irrigation and environmental quality at the hydrological basin level - I. Irrigation quality. *Agricultural Water Management*, Vol.70, No.3, (December 2004), 195-209, ISSN 0378-3774.

Causapé, J.; Quílez, D. & Aragüés, R. (2004c). Assessment of irrigation and environmental quality at the hydrological basin level - II. Salt and nitrate loads in irrigation return flows. *Agricultural Water Management*, Vol.70, No.3, (December 2004), 211-228, ISSN 0378-3774.

Causapé, J.; Quílez, D. & Aragüés, R. (2006). Irrigation efficiency and quality of irrigation return flows in the Ebro River Basin: An overview. *Environmental Monitoring and Assessment*, Vol.117, No.1-3, (June 2006), 451-461, ISSN 0167-6369.

Cavero, J.; Beltrán, A. & Aragüés, R. (2003). Nitrate exported in drainage waters of two sprinklerirrigated watersheds. *Journal of Environmental Quality*, Vol.32, No.3, (May 2003), 916-926, ISSN 0047-2425.

CHE (October 2010). Geoportal SITEbro, Available from http://iber.chebro.es/geoportal/index.htm

Christen, E.W.; Ayars, J.E. & Hornbuckle, J.W. (2001). Subsurface drainage design and management in irrigated areas of Australia. *Irrigation Science*, Vol.21 (December 2001), 35-43, ISSN 0342-7188.

Clemmens, A.J. & Dedrick, A. R. (1994). Irrigation techniques and evaluations. In: *Advanced Series in Agricultural Sciences*. Tanji, K.K. & Yaron, B. (Eds.), 64-103, Springer-Verlag, ISBN 3-540-57309-7, Berlin.

Cui, Z., Zhang, F., Chen, X., Dou, Z., Li, J. (2010). In-season nitrogen management strategy for winter wheat: Maximizing yields, minimizing environmental impact in an over-fertilization context. *Field Crops Research*, Vol.116 (March 2010), 140-146, ISSN 0378-4290.

Custodio, E. & Llamas, M. (1983). *Hidrología Subterránea*. Ediciones Omega, ISBN 84-282-0446-2, Barcelona.

Dechmi, F.; Playán, E.; Cavero, J.; Faci, J.M. & Martínez-Cob, A. (2003). Wind effects on solid set sprinkler irrigation depth and yield of maize (Zea mays). *Irrigation Science*, Vol.22, No.2, (September 2003), 67-77, ISSN 0342-7188.

Diaz, R.J. (2001). Overview of hypoxia around the world. *Journal of Environmental Quality*, Vol.30 (March 2001), 275-281, ISSN 0047-2425.

Duncan, R. A.; Bethune, M. G.; Thayalakumaran, T.; Christen, E. W. & McMahon, T. A. (2008). Management of salt mobilisation in the irrigated landscape. A review of selected irrigation regions. *Journal of Hydrology*, Vol.351, No.1-2, (March 2008), 238-252, ISSN 0022-1694.

EU (1991). *Council Directive 91/676/EEC of 12 December 1991 concerning the protection of waters against pollution caused by nitrates from agricultural sources*, European Council.

FAO (2002). *World agriculture towards 2015/2030. An FAO perspective*. Food and Agriculture Organization of the United Nations. ISBN 92-5-304761-5, Roma, Italy.

FAO (2003). *Unlocking the water potential of agriculture*. Food and Agriculture Organization of the United Nations, ISBN 92-5-104911-4, Roma, Italy.

GA (November 2009a). Oficina del Regante, Datos meteorológicos, Available from http://oficinaregante.aragon.es

GA (November 2009b). Anuario estadístico agrario, Available from http://www.aragon.es/DepartamentosOrganismosPublicos/Departamentos/Agri culturaAlimentacion/AreasTematicas/EstadisticasAgrarias?channelSelected=1cfbc 8548b73a210VgnVCM100000450a15acRCRD

García-Garizábal, I., Causapé, J., Abrahão, R. (2009). Evolution of the efficiency and agro-environmental impact of a traditional irrigation land in the middle Ebro Valley (2001-2007). *Spanish Journal of Agricultural Research*, Vol.7 (June 2009), 465-473, ISSN 1695-971X.

García-Garizábal, I. & Causapé, J. (2010). Influence of irrigation water management on the quantity and quality of irrigation return flows. *Journal of Hydrology*, Vol.385, No.1-4, (May 2010), 36-43, ISSN 0022-1694.

Gheysari, M., Mirlatifi, S.M., Homaee, M., Asadi, M.E., Hoogenboom, G. (2009). Nitrate leaching in a silage maize field under different irrigation and nitrogen fertilizer rates. *Agricultural Water Management*, Vol.96 (June 2009), 946-954, ISSN 0378-3774.

Isidoro, D.; Quílez, D. & Aragüés, R. (2004). Water balance and irrigation performance analysis: La Violada irrigation district (Spain) as a case study. *Agricultural Water Management*, Vol.64, No.2, (January 2004), 123-142, ISSN 0378-3774.

ITGE (1985). *Investigación de los recursos hidráulicos totales de la cuenca del río Arba*. Instituto Tecnológico Geominero de España, Spain.

ITGE (1995). *Informe complementario del mapa geológico de Luna. Hidrogeología de la hoja de Luna (27-11)*, Instituto Tecnológico Geominero de España, Spain.

Lecina, S.; Playan, E.; Isidoro, D.; Dechmi, F.; Causapé, J. & Faci, J. M. (2005). Irrigation evaluation and simulation at the irrigation District V of Bardenas (Spain). *Agricultural Water Management*, Vol.73, No.3, (May 2005), 223-245, ISSN 0378-3774.

Moreno, F., Cayuela, J.A., Fernández, J.E., Fernández-Boy, E., Murillo, J.M., Cabrera, F. (1996). Water balance and nitrate leaching in an irrigated maize crop in SW Spain. *Agricultural Water Management*, Vol.32 (November 1996), 71-83, ISSN 0378-3774.

Martínez-Cob, A. (November 2010). Revisión de las necesidades hídricas netas de los cultivos de la cuenca del Ebro, Available from http://digital.csic.es/bitstream/10261/15896/1/NecHidrCHE04_T1_Mem.pdf

Orús, F. & Sin, E. (2006). *El balance del nitrógeno en la agricultura. Fertilización nitrogenada. Guía de actualización*. Gobierno de Aragón, ISSN 1137-1730, Zaragoza, Aragón, Spain.

Playán, E., Salvador, R., Faci, J.M., Zapata, N., Martínez-Cob, A., Sánchez, I. 2005. Day and night wind drift and evaporation losses in sprinkler solid-sets and moving laterals. *Agricultural Water Management*, Vol.76, No.3, (August 2005), 139-159, ISSN 0378-3774.

Roman, R.; Caballero, R. & Bustos, A. (1999). Field water drainage under traditional and improved irrigation schedules for corn in central Spain. *Soil Science Society of America Journal*, Vol.63, No. 6, (November-December 1999), 1811-1817. ISSN 0361-5995.

Scavia, D. & Bricker, S.B. (2006). Coastal eutrophication assessment in the United States. *Biogeochemistry*, Vol.79(May 2006), 187-208, ISSN 0168-2563.

Tanji, K.K. & Kielen, N.C. (2002). *Agricultural drainage water management in arid and semi-arid areas*. FAO irrigation and drainage paper nº 61, ISBN 92-5-104839-8, Roma, Italy.

Thayalakumaran, T., Bethune, M.G., McMahon, T.A. (2007). Achieving a salt balance - Should it be a management objective?. *Agricultural Water Management*, Vol.92 (August 2007), 1-12, ISSN 0378-3774.

Urdanoz, V.; Amezqueta, E.; Clavería, I.; Ochoa, V. & Aragües, R. (2008). Mobile and georeferenced electromagnetic sensors and applications for salinity assessment. *Spanish Journal of Agricultural Research*, Vol.6, No.3, (September 2008), 469-478, ISSN 1695-971X.

SIAS (November 2009). Sistema de Información del Agua Subterránea, Available from http://www.igme.es/internet/ServiciosMapas/siasespana/sias-es.html

Soil Survey Staff (1992). *Keys to soil taxonomy*. Pocahontas Press, ISBN 0-936015-39-X, Blacksburg, Virginia, USA.

Tedeschi, A.; Beltrán, A. & Aragüés, R. (2001). Irrigation management and hydrosalinity balance in a semi-arid area of the middle Ebro river basin (Spain). *Agricultural Water Management*, Vol.49, No.1, (July 2001), 31-50, ISSN 0378-3774.

Tanji, K.K. & Kielen, N.C. (2002). *Agricultural drainage water management in arid and semi-arid areas*. Food and Agriculture Organization of the United Nations, ISBN 92-5-104219-5, Roma, Italy

Wahl, T. (2000). *Winflume user's manual*. United States Department of the Interior, USA.

Wang, B. (2006). Cultural eutrophication in the Changjiang (Yangtze River) plume: History and perspective. *Estuarine Coastal and Shelf Science*, Vol.69 (September 2006), 471-477, ISSN 0272-7714.

WHO (2004). *Guidelines for drinking-water quality. Vol.1: Recommendations*. World Health Organization, ISBN 92-4-154638-7, Geneva, Italy.

Zalidis, G.; Dimitriads, X.; Antonopoulos, A. & Geraki, A. (1997). Estimation of a network irrigation efficiency to cope with reduced water supply. *Irrigation Drainage Systems*, Vol.11, No.4, (November 1997), 337-345, ISSN 0168-6291.

Zapata, N., Playan, E., Martinez-Cob, A., Sanchez, I., Faci, J.M., Lecina, S. 2007. From on-farm solid-set sprinkler irrigation design to collective irrigation network design in windy areas. Agricultural Water Management 87 (2): 187-199.

Zapata, N., Playán, E., Skhiri, A., Burguete, J. (2009). Simulation of a collective solidset sprinkler irrigation controller for optimum water productivity. Journal of Irrigation and Drainage Engineering 135 (1): 13–24.

Comparing Water Performance by Two Different Surface Irrigation Methods

Francisco Mojarro Dávila[1], Carlos Francisco Bautista Capetillo[1],
José Gumaro Ortiz Valdez[1] and Ernesto Vázquez Fernández[2]
[1]Universidad Autónoma de Zacatecas
[2]Universidad Nacional Autónoma de México
México

1. Introduction

The crop optimal growth demands adequate water supply. When rainfall is not sufficient in a region to satisfy crop water requirements it has to be complemented with irrigation water in order to replace evapotranspiration losses occurred in a specific period so that quality and yield are not affected (Brouwer et al., 1988; Ojeda et al., 2007). A field receives irrigation water using pressurized systems or by water flows from its available energy, basically. This last case is called surface irrigation and includes a large variety of irrigation systems sharing a common characteristic: water is applied on the soil surface and is distributed along the field by gravity. This fact marks the importance to analyze infiltration process and water retention capacity of soils as the most important physical properties involved in water dynamics around roots zone (Playán, 2008; Walker & Skogerboe, 1987).

Surface irrigation continues being the most used irrigation system in the world even thought its efficiency range between 30% and 50% (Rosano-Méndez et al., 2001; Sió et al., 2002); nevertheless Hsiao et al. (2007) discussed some works (Erie & Dedrick, 1979; Howell 2003) which conclude that application efficiency can be higher than 80% if surface irrigation is practiced well under the right conditions. The low irrigation efficiency combined with the decreasing in water availability for irrigation due to severe and extended droughts as well as the great competition that has been occurring among all users (such as residential users, industries, and farmers) that started twenty-five years ago, it raises the opportunity so that surface irrigation agriculture makes a more rational water use because it shows two important advantages regarding to pressurized irrigation: 1) field does not requires equipment, and 2) pumping is not necessary at field level; so equipment and pumping energy costs are lower. Nevertheless to provide volumes to be used by crops with minor water losses can bring environmental implications (less runoff, less volume for aquifer recharge; for example).

During water movement into soil profile hydrological processes of different nature appear; for this reason surface irrigation is divided in phases to separate them. In each one there are peculiarities that allow obtaining some characteristic times (Walker, 1989; Khatri, 2007): *a) Starting time*, water begins to flows in the field –border, basin or furrow. *b) Time of advance*, water completely covers the basin or border, or water reaches the downstream end of a furrow. *c) Time of cut off*, water stops flowing into the irrigated field. *d) Time of depletion*, when a part of the basin, border or furrow becomes uncovered by water once the water has fully infiltrated or has moved to lower areas of the field. *e) Time of recession*, water can no longer be seen over the field. The difference between time of advance and time of recession is known as opportunity time. During this period occurs the infiltration process. Surface irrigation phases (Figure 1) are defined as: *a) advance phase*, water flows in non-uniform and spatially varied regime, the discharge decrease downstream for the infiltration process in porous media consequently. In this phase water is covering border, basin or furrow. *b) Filling up phase*, once water reaches the end of the field, the discharge remains during sufficient time to apply the water table required by crop. In this phase the water is filling up the soil pores. *c) Depletion phase*, water is cut off causing a gradual diminution in water depth; this phase ends when the water has been totally infiltrated in a portion of the field. *d) Recession phase*, water uncovers the field surface completely as a wave moving at the same direction of flow.

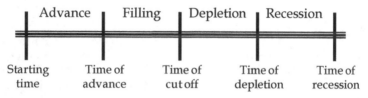

Fig. 1. Irrigation times and phases (adapted from Playán, 2008)

Border, basin and furrows are the most common methods in surface irrigation. Four different variants to transport water into the fields using furrows are being developed: *a)* continuous-flow irrigation (CFI), *b)* intermittent (pulses) irrigation (II), *c)* cut-back irrigation (CBI), and *d)* increased-discharge irrigation (IDI) (Vázquez, 2001). In Mexico 5.4 million hectares are irrigated by some surface irrigation method. Farmers that apply water in furrows as transportation media are using CFI principally so water discharge never is cutting along all longitude of furrow although it arrives at the end of it. Its mean efficiency is around 59% (Alexander-Frezieres, 2001); even though it is low some improvements on the irrigation settings are already taken place; efficiency in farming fields continues been low. Montiel-Gutiérrez (2003) conducted field measurements in an irrigation zone; whose results showed an application mean efficiency of 57% and 39%. For that reason, it is necessary to prove alternative methods to that of continuous flow irrigation. According to Vazquez et al. (2003) an option is IDI. This option required a previous improvement in techniques of field irrigation, by means of installation of gated pipe, which already were used in several regions of the country. The IDI consists in applying water initially as the total volume flowing through the gated pipes to all furrows in a battery; then, once the water front approaches one quarter of the furrow length, half of the gates are closed, this causes the duplication of the inflow in the furrows with the gates open. Once irrigation is completed in the first half of the battery, the total flow of piping is applied to the other half of the battery which is temporarily interrupted; the previously opened gates are to be closed and opened the ones that were closed, to achieve an increment in flow. The irrigation of that furrows (second half) has a discontinuous irrigation, with the double volume compared to that the initial flow (Ortiz, 2005). On summary, this technique is the opposite to that of "cut-back" proposed by Humphreys (1978).

The purposes of this work was to comparing water performance by CFI and IDI methods in blocked-end furrows for maize crop in two seasons (2004-2005) during spring-summer as well as analyze furrow irrigation variables (inflow discharge, water table, and time of irrigation cutoff) and their relation to performance irrigation indicators of water use: efficiency, irrigation efficiency, water productivity, and crop production. Herein, a computer program was used to simulate the furrow irrigation process (Vázquez, 2001).

2. Study zone description

2.1 General characteristics of the study area

The experimental plot was located at the experimental station of the National Research Institute for Forestry, Agriculture and Livestock (INIFAP), situated to the northwest of the city of Zacatecas, Mexico with geographic coordinates: North Latitude 22° 54' 22.3" and Longitude West 102° 39' 50.3" and an average elevation above sea level to 2200 m (Figure 2).

Fig. 2. Location study area at INIFAP experimental station

The climate is characterized as semi-arid, where average annual evaporation exceeds in 2,000 mm to average annual precipitation, with summer rains and very scarcity in the rest of the year, average annual precipitation is 419.8 mm, average annual reference evapotranspiration (ET₀) is around 1,490 mm and average temperature range between 12 to 18 °C. Within the experimental station there is an automated weather station from which data was collected for this study. Monthly rainfall, temperature and reference evapotranspiration recorded for corn grow cycle in the two years of study is presented in Figure 3.

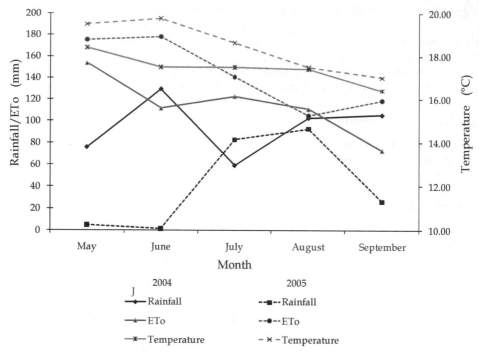

Fig. 3. Monthly rainfall, ET_0 and temperature recorded at the INIFAP experimental station

2.2 Soil physics characterization

From the experimental plot soil samples were taken at six random points to the depth of 0-60 cm and in the laboratory the following soil characteristics were identified (Table 1). The inflow and outflow method (USDA, 1956) were used to determine soil basic infiltration rate. Three 90 degree triangle flumes were previously calibrated in the laboratory. The flumes were installed on a furrow at distances of 50, 100 and 150 from inlet point and water levels were recorded every 5 minutes. The inflow into the furrow (0.75 m spacing) was delivered from the field by using a gated pipe and the inflow was maintained constant during all time. The soil basic infiltration rate was 1.1 cm h^{-1}.

2.3 Maize crop

In the two years of study the hybrid H-311 was selected which is a hybrid semi-late with white grain. Its height is 2.70 m, the stems are strong and time to maturity is 150 days. Economic yield is from 6,500 to 8,500 kg of dry grain per hectare (Luna & Gutierrez, 1997). Planting took place on April 10 of 2004 and in 2005 on 15 April, with a density of 65,000 plants per hectare and the fertilization was N=200 kg ha^{-1}, P=80 kg ha^{-1}, and K=00 kg ha^{-1} in both years. To estimate crop evapotranspiration, historical average weather data of temperature and precipitation from the INIFAP weather station were used. These values were used to run PIREZ software (Integrated Irrigation Project for the State of Zacatecas) (Mojarro et al., 2004); resulting a crop evapotranspiration around 50.8 cm for corn season (sowing to harvest).

Sample	1	2	3	4	5	6
Texture	Silty loam	Silty loam	Silt	Silty loam	Silty loam	Silty Loam
Sand (%)	30.44	36.24	36.24	37.88	41.88	43.88
Silt (%)	34.00	32.00	38.00	32.00	30.00	28.00
Clay (%)	35.00	31.76	25.76	30.12	28.12	38.12
Hydrodynamic constants:						
Field Capacity	23.9%					
Permanent Wilting Point	12.8%					
Bulk Density	1.27 g/cm³					

Table 1. Texture and hydrodynamic constants of soil samples of experimental field

3. Methodology

3.1 Experimental plot

Topographic survey of the experimental plot was performed with a total station Sokkia brand SET Model 610. On the plot a grid of 10 x 10 m was formed to get the lay of the land and after that land plane was carried out by an electronic equipment of leveling laserplane. The dimensions of the experimental plot were 20 m wide by 198 m long, with a land surface slope of 0.25%. In this plot the experimental work was performed for evaluation of irrigation method with increased-discharge (IDI) in comparison with the traditional irrigation method in which the inflow was constant (CFI). There were 12 blocked-end furrows where the IDI was established and 12 blocked-end furrows where the CFI was established.

3.2 Field experiment management

A gated pipe of 6 inches in diameter was used for water application. This irrigation system is very common among irrigation farmers in the study area, Figure 4 shows the characteristics of the experimental plot and for the irrigation management was as follows:

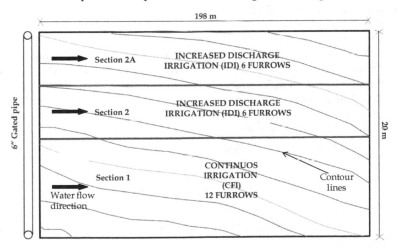

Fig. 4. Topographical diagram for experimental plot

1) there was an auxiliary plot with 12 furrows where the inflow for each furrow was calibrated and fixed; 2) once this happened, the gates in the section 1 (Figure 4) were opened until irrigation time was achieved; 3) 12 gates of section 2 and section 2A were opened; 4) when the water front reached 50 m, six gates were closed (Section 2A) and then for the other six gates, the inflow per furrow was increased two fold, until the irrigation time was achieved; 5) once this happened, the gates in section 2A were opened until the irrigation time was completed. The consumption time in the operation of 12 furrows for IDI was less than three minutes for each irrigation event.

3.3 Simulation models
Mathematical simulation models are a useful tool in the design and / or correction of inflow, the slope and the roughness of surface irrigation. However these models require knowledge of the function of the soil infiltration, but its determination in the field is not easy due to the spatial variability of soils (Rendón, et al. 1995). Moreover the advance of water on the furrow surface is dominated by the forces of gravity and is expressed by Saint-Venant equations, which represent the total hydrodynamic phenomenon (Vázquez, 1996). The simulation model in blocked-end furrow proposed by Vazquez (RICIG) (2001) has the attributes to simulate CFI (traditional) and IDI. The RICIG uses the Green and Ampt equation considering the initial soil moisture and uses the wetted perimeter to calculate the infiltration; in addition to considering the water flow on a furrow surface is transitional and gradually varied because the water infiltrates into the soil as it moves toward the end of the furrow. RICIG model includes equations that play this type of flow which are the continuity and momentum, both known as the Saint-Venant equations. Vázquez (1996) comments that these equations have as unknowns the inflow and the depth of water in different sections, and it is assumed that the channel or furrow has a prismatic form which does not change all along the furrow, and the soil is homogeneous this means that the hydraulic conductivity is constant along the furrow.

3.4 Variables measured in the field
3.4.1 Soil moisture content
One of the most important effects on the irrigated agriculture is to fully satisfy the soil moisture in the root zone of the crop. The soil water content should be measured periodically to determine when to apply the next irrigation and how much water should be applied. With these purposes in 2004 and 2005, soil moisture content was measured once a week and before and after the irrigation event, to the depths of 0-15, 15-30 and 30-45 cm. The gravimetric method was used; samples were taken with the Vehimeyer auger recommended by the EPA (2000). To calculate the soil moisture content was used the equation 1 and equation 2 was used to calculate water table.

$$W_i\,(\%) = \left(\frac{W_{ws} - W_{ds}}{W_{ds}} \right) 100 \tag{1}$$

Where: Wi is the moisture content (%);Wws is the weight of wet soil (g); and Wds is the weight of dry soil (g).
Water table applied for each irrigation was estimated according to equation (2)

$$Z_m = P_r \frac{\gamma_i}{\gamma}\left[\left(WC_{fc} - WC_i\right)/100\right] \qquad (2)$$

Where: Z_m is the water table to implement (cm); P_r is the root depth (cm); γ_i is the volumetric weight of the soil (kg m-3); γ is the specific weight of water (1000 kg m-3); WC_{fc} is moisture content at Field Capacity (%) of dry weight; WC_i is the residual moisture content before irrigation (%) of dry weight.

3.4.2 Cross-sectional surface flow area
One of the input variables in the RICIG model is the cross-sectional geometry of furrows, this was done only in the year 2004 with a profilometer (Picture 1) designed by Ortiz (2005). A total of 120 cross-sectional geometry of furrow were made during the course of the experiment mainly before sowing irrigation, after agricultural practices and after each event of irrigation.

Picture 1. Furrow profilometer for determining cross-sectional area

3.4.3 Advance of water over furrow surface
In 2004, the advance of the water was determined at each irrigation event. Along the furrows 10 marking points were placed (20 m apart) to measure the time it takes water flow to reach those points (Picture 2).

3.5 Application efficiency and irrigation water use efficiency and productivity
The irrigation efficiency is clearly influenced by two factors: 1) the amount of water used by the crop for water applied in irrigation, and 2) distributions in the field of applied water.

These factors affect the cost efficiency of irrigation, irrigation design, and most important, in some cases, productivity of crops. Efficiency in water use has been the most widely used parameter to describe the efficiency of irrigation in terms of crop yield (Howell, 2002). For the years 2004 and 2005 the following indicators of irrigation efficiency and water productivity were used. Application efficiency (E_a) was evaluated considering the methodology proposed by Rendón et al. (2007). E_a is defined as the amount of water that is available to crops in relation to that applied to the plot and was calculated as:

$$E_a = \frac{V_i}{V_a} \tag{3}$$

Where Va is the total volume applied to the plot (m³); Vi is the irrigated volume usable by the crops (m³). The applied volume (Va) is defined as:

$$V_a = Q_e\, t_r \tag{4}$$

Qe is the inflow applied to the plot (m³ s⁻¹) and t_r is the time of irrigation (s). The volume is still available (V_e) to plants can be defined as:

$$V_e = V_a - V_r \tag{5}$$

Vr is the infiltrated volume beyond the root zone (m³). To estimate Vr it is necessary to determine the soil moisture in the soil profile along the furrow before and after each irrigation event.

Picture 2. View of the points where the time water flow was recorded

The distribution uniformity is defined as a measure of the uniformity with which irrigation water is distributed to different areas in a field. The distribution uniformity of the infiltrate depth was estimated by the distribution coefficient of Burt et al. (1997) using equation.

$$DU=\frac{Z_{min}}{Z} \qquad (6)$$

Where Z_{min} is the minimum infiltration depth in a quarter of the total length of the furrow (cm) and Z is the average of the infiltrated depth (cm). Water use efficiency (WUE) is the ratio between economic yield (Y_e) and crop evapotranspiration (ET_C) (Howell, 2002) for the study region $ET_C = 50.8$ cm, as fallows.

$$WUE=\frac{Y_e}{ET_c} \qquad (7)$$

The irrigation water use efficiency (IWUE) is the ratio of the difference of Ye and crop economic yield under rain-fed conditions (Yt) (no irrigation is applied and Yt= 550 kg ha-1) and the applied water table (Zm) (Howell, 2002) using the next equation.

$$IWUE=\frac{Y_e-Y_t}{Z_m} \qquad (8)$$

Water productivity (WP) is the quotient obtained by dividing economic yield of the crop (Y_e) and the volume applied (Va), as fallows.

$$WP=\frac{Y_e}{V_a} \qquad (9)$$

4. Results

In order to define the inflow discharge (Q) to be applied in the design of CFI and IDI irrigation methods, nine irrigation tests were conducted in which discharge varied from 0.9 to 2.2 l s-1. Figure 5 shows water advance curves and as expected they are different for each Q used (similar results were found in several works, for example Bassett et al. 1983). In addition they show that the lower Q has a lower water speed therefore water had more opportunity time to infiltrate, and this creates a larger water table at the beginning of the furrow and a poor distribution of soil moisture along the furrow (data not presented).

Also another test was conducted with data from Table 1, Q=2.2 l s-1, and using the CFI proceeded to the simulation with the RICIG model and realization under the field condition of the irrigation in blocked-end furrow. Figure 6 illustrates that the water advance curves obtained with the simulator and with the observed data from the field are very similar. So it follows that the mathematical model (RICIG) represents the physical phenomenon of surface irrigation in blocked-end furrows.

4.1 Applied irrigations
The data obtained from tests carried out, furrow geometry, and data from Table 2 pre-sowing irrigation was simulated and applied in the field.

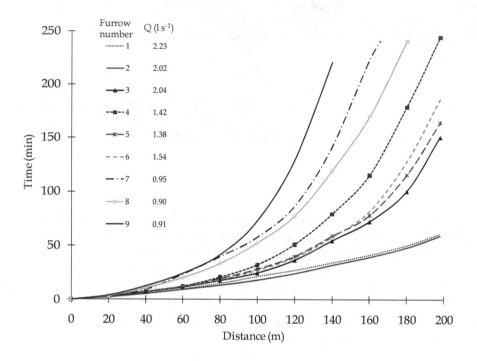

Fig. 5. Family water advance curves

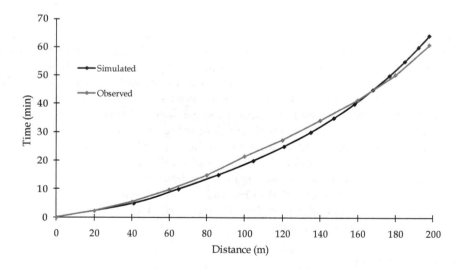

Fig. 6. Water advance curves. Pre-sowing irrigation for CFI (2004)

Variable	CFI	IDI
Q (l s⁻¹)	2.4	2.4
t_r (min)	65	31
Z_m (cm)	6.0	6.0

Table 2. Data required for simulating and applying the pre-sowing irrigation (2004)

Figures 7 and 8 present simulated and observed water advance curves for continuous-flow irrigation and increased-discharge irrigation respectively. In those figures it shows that in the case of CFI curves exhibit a rising exponential of the traditional way but instead the points for the IDI approximately fit a straight line, which explains a considerable reduction

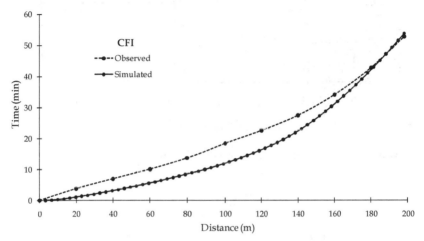

Fig. 7. Water advance curves for CFI. Observed values were monitored in 2004

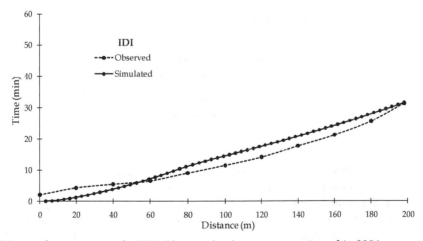

Fig. 8. Water advance curves for IDI. Observed values were monitored in 2004

in time of the advance phase. Figure 9 shows the inflow hydrograph for the first 6 furrows irrigated with increased inflow and without interrupting water supply in the furrow (see Figure 4, section 2). Figure 10 shows the inflow hydrograph for the other six furrows irrigated in which the inflow was interrupted (Figure 4, section 2A) when the advancing front of the water reaches a quarter of the length of the furrow with a break time (tint) at 7.3 min, subsequently returned to this part when the time of irrigation (tr= 28.4 min) was completed in the first 6 furrows.

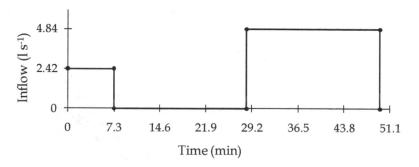

Fig. 9. Inflow hydrograph for IDI. First 6 irrigated furrows in the furrows battery

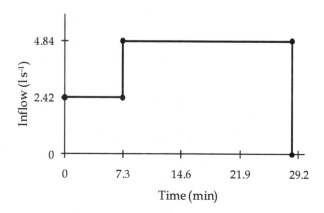

Fig. 10. Inflow hydrograph for IDI. Second 6 irrigated furrows in the furrows battery

According to the hydrographs (Figures 9 & 10) and the information on Table 3, it shows that for the IDI the irrigation time was 49.2 min and for CFI it was 68.9 min. Table 3 presents the results of simulations and field observations for CFI and IDI. According to this information, the irrigation time was about 20 min higher in CFI than in the IDI. The distribution uniformity (DU) was 24% higher in the IDI than the CFI. The application efficiency (Ea) was higher 16.6% in the IDI. The applied volume (Va) was higher in CFI than in the IDI.

Irrigation Method	Q (l s⁻¹)	t_r (min)	Z_m (cm)	V_a (m³)	DU (%)	E_a (%)
CFI (simulated)	2.4	72.0	6.1	8.95		66.6
CFI (observed)	2.4	68.9	6.8	10.1	69.0	
IDI (simulated)	2.4	34.0	6.1	7.15		83.2
IDI (observed)	2.4	49.2	4.1	6.15	90.3	

Table 3. Simulated and observed data in the second irrigation (2004)

The profiles of simulated and observed water infiltration table found for the pre-sowing irrigation are presented in figures 11 and 12. There are some differences between observed and simulated values for both methods; these differences can be explained in part because the observed field values were performed 48 hours after irrigation, however minor differences are observed in IDI. If the observed values for both methods are compared, it is showed that the water infiltration table in CFI is greater at the beginning of furrows than in the IDI and it basically is due to the lower speed of water in the furrows with CFI method, and the uniformity of distribution in IDI is better than the CFI, which allows a better crop water use.

Table 4 displays that the mean of irrigation time (tr) for CFI is 75.3 minutes while for the IDI is only 58.7 min., which represents a time saving of 16.6 minutes per irrigation applied. The total applied water tables (Zm) are 47.2 and 38.9 cm. respectively, thus saving water in the IDI is 830 m³ per hectare. The efficiency of water used is tied to the ability to achieve and understand the integrated system water-soil-plant-atmosphere, which is the basis for making decisions on when and how much water to be applied. In the experimental plot during the course of the experiment the soil moisture content was measured once a week and before irrigation, and 48 hours thereafter.

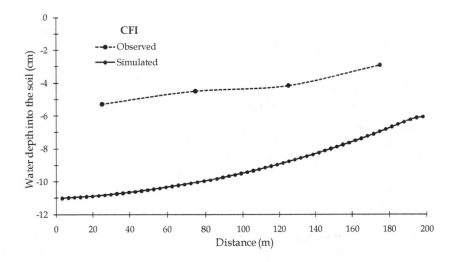

Fig. 11. Infiltration table along furrow for CFI. Observed values were monitored in 2004

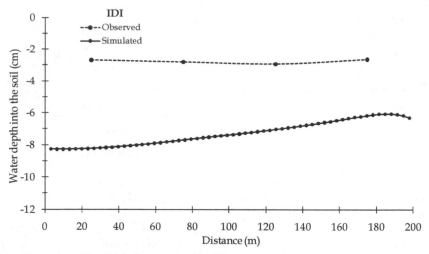

Fig. 12. Infiltration table along furrow for IDI. Observed values were monitored in 2004

Irrigation	Date	Q (l s⁻¹)	t_{inc} (min)	t_r (min)	V_a (m³)	Z_m (cm)	t_r (min)	V_a (m³)	Z_m (cm)
				Increased Irrigation (IDI)			**Continuous Irrigation (CFI)**		
Pre-sowing	03-15	2.5				9.0			9.0
2	05-08	2.4	7.3	42.2	6.1	4.1	69.8	10.1	6.8
3	05-22	2.5	6.6	54.5	8.4	5.6	70.2	10.8	6.8
4	06-10	2.5	6.5	49.3	7.6	5.1	50.9	7.8	5.2
5	07-10	2.1	7.7	50.6	7.6	5.1	60.0	7.6	5.1
6	07-16	1.6	7.7	50.6	6.4	4.3	60.9	6.0	4.0
7	07-27	1.2	12.0	67.3	4.9	3.2	90.3	6.5	4.4
8	08-20	1.0	15.0	96.0	6.2	4.2	125.0	8.0	5.4
Total Mean				58.7		38.9	75.3	77.6	47.2

Table 4. Results of irrigation application on corn

Figure 13 was developed with average values of soil moisture content of all irrigation events. This figure shows that the soil moisture content 48 hours after irrigation is higher in irrigated furrows where the CFI was applied than it was in IDI, there is a clear separation between the curves at the beginning of the furrows and then they decrease as the furrows near to the end. In the case of soil moisture content before irrigation there is a slight difference between the curves.

4.2 Soil moisture content measurements and evaluation of uniform distribution
With the sampling of soil moisture the water tables were obtained as well as the distribution uniformity was estimated (DU). Tables 5 and 6 present the values of water tables and DU for CFI and IDI methods respectively. Tables 5 and 6 show that the DU mean for IDI was 89.6%

and 75.6% for CFI this means that the uniformity of distribution of Zm were higher in IDI 14% more than in the CFI. Therefore there is an evidence to encourage the IDI irrigation method. The application efficiency was estimated for CFI and IDI; the values were 66.6% and 83.2% respectively. According to these and previous results, there is strong evidence that IDI irrigation method is more efficient than the CFI irrigation method (Tables 3, 5 and 6 and Figures 11, 12, and 13).

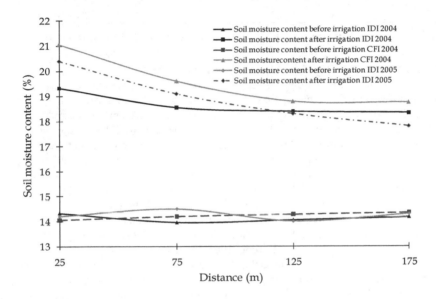

Fig. 13. Average soil moisture content before and after irrigation events (2004 & 2005)

Irrigation number	Distance (m)				Water table (cm)		DU
	25	75	125	175	Mean	Minimum	(%)
1	6.5	4.4	3.8	3.5	4.6	3.5	77.3
2	6.0	4.9	4.2	3.4	4.6	3.4	73.8
3	6.2	5.6	3.5	3.5	4.7	3.5	74.8
4	5.1	2.8	2.7	2.7	3.5	2.7	77.5
5	3.9	3.2	2.9	2.9	3.3	2.9	86.8
6	4.2	3.8	4.0	4.0	3.6	2.3	63.6
						DU (mean)	75.6

Table 5. Water table at different distances of the furrows and DU for CFI method

| Irrigation | Distance (m) | | | | Water table (cm) | | DU |
number	25	75	125	175	Mean	Minimum	(%)
1	3.4	3.0	3.0	2.9	3.0	2.9	94.4
2	4.1	4.0	3.4	4.1	3.9	3.4	87.2
3	4.4	3.6	3.3	3.1	3.6	3.1	85.4
4	3.7	3.2	2.8	2.9	3.2	2.8	90.0
5	2.9	3.1	3.1	2.7	3.0	2.7	92.4
6	3.7	3.2	3.1	2.8	3.2	2.8	88.4
						DU (mean)	89.6

Table 6. Water table at different distances of the furrows and DU for IDI method

4.3 Water productivity, irrigation efficiency, and water use

The harvest took place on November 11 in 2004 and 2005 on November 5. Thirty two points were set sampling for IDI and others for CFI, where the harvested area per sampling point was 5.2 m2. Table 7 shows the found economic yields (Y_t) for different irrigation methods and years. To determine the average water productivity performance the economic yield was divided between the volume of water applied and for the CFI water productivity was 1.83 kg m^{-3} (2004), while for the IDI in the year 2004 was 2.34 kg m^{-3} and in the year 2005 of 1.93 kg m^{-3} (Table 7), these differences between years are explained by the environmental conditions that were observed, recorded rainfall during the growing season was higher in 2004 with more 325 mm than 2005 and on the other side, the pan evaporation observed during 2005 was 315.05 mm more than in 2004 for the same period. Grain yield of maize was acceptable because it exceeded the state average (4.04 T ha-1). From the above evidences, water productivity with the IDI (2004 and 2005) is 1.28 and 0.13 times the CFI (2004) and 2.54 and 2.1 higher than that reported by Chavez (2003). In relation to the WUE, and IWUE, those values were higher in IDI than in the CFI (Table 7). In addition to, it is clear that the IWUE, in general varies from year to year and with the use of better technology, such as in the work of Chavez (2003) where the design of irrigation was not carried out.

	IDI (2004)	CFI (2004)	IDI (2005)	CFI (Chávez, 2003)
Crop	Corn	Corn	Corn	Corn
Land leveling	Yes	Yes	Yes	Yes
Irrigation design	Yes	Yes	Yes	No
Dose of Fertilizer	200-80-00	200-80-00	200-80-00	235-60-00
Irrigation events	8	8	11	8
V_a (m^3 ha^{-1})	3,892.5	4,720.0	5,980.0	6,511.0
Y_t (kg ha^{-1})	9,119.0	8,641.0	11,585.0	6,000.0
WP (kg m^{-3})	2.34	1.83	1.93	0.92
WUE	179.5	170.0	228.0	128.1
IWUE	2.20	1.71	1.84	0.83

Table 7. Indicators of irrigation water efficiency

5. Conclusions

In the IDI method the phase of water advance is reduced due to the inflow discharge was increased and therefore the uniformity of distribution of the furrows was increased, being reflected in the reduced irrigation time, and water table and increased crop production due to better distribution uniformity of the water table. The WP average in the IDI (2004 and 2005) is 2.13 kg m-3, while the CFI is 1.83 kg m-3, which represents 25% more productivity. The efficiency of water distribution in the root zone of the crop during the growing season for two years of study in the IDI was higher by 15.6% than in CFI which means that the crop had better conditions of soil moisture for a higher value of WP. The use of irrigation method IDI compared with CFI offers clear advantages for maize production, since WUE by the crop is improved by an average of 27%, the IWUE is increased by 16%, and the irrigation time is reduced by 23 min per irrigation. Therefore there is an evidence to encourage the IDI irrigation method. Water use efficiency, irrigation water use efficiency, and distribution uniformity are the performance irrigation parameters that seem to be correlated with furrow irrigation variables (water table, inflow discharge, volume applied and time of irrigation) it was observed by Holzapfel et al., (2010). They are also the parameters that thus could have a relationship between crop productivity and production and the irrigation variables. Therefore, IDI irrigation method is recommended to be used for establish good irrigation practices. The future research should be aimed at determining the optimal inflow discharge on the IDI, for different characteristics of blocked-end furrow such as length, slope and roughness.

6. Notation

The following symbols are using in this chapter
CBI = cut-back irrigation
CFI = continuous-flow irrigation
DU = distribution uniformity
Ea = application efficiency
ETc = crop evapotranspiration
ETo = reference evapotranspiration
IDI = increased-discharge irrigation
II = Intermittent irrigation
IWUE = irrigation water use efficiency
P_r = root depth
Q = inflow discharge
t_{inc} = interrupted time
t_r = irrigation time
Va = applied volume
Ve = volume still available
Vi = volume usable by the crops
Vr = infiltrated volume
WC_{fc} = field capacity
Wds = weight of soil dry
W_i = soil moisture content

WP = water productivity
WUE = water use efficiency
Wws = weight of soil wet
Y_e = economic yield
Y_t = economic yield under rain-fed conditions
Z = average of the infiltrated depth
Z_m = water table
Zmin = minimum infiltration depth in a quarter of the total length of the furrow
γ = specific weight of water
γi volumetric weight of the soil

7. Acknowledgement

The authors are in complete gratitude to Jose Gumaro Ortiz Valdez for using the results of his master's thesis for the enforcement of this chapter.

8. References

Alexander-Frezieres, J. (2001). Conservación de la Infraestructura Hidroagrícola en las Unidades de Riego en México, *XI Congreso Nacional de Irrigación*, Guanajuato, México

Bassett, L. D.; Frangmeier, D. D.; & Strelkoff, T. (1983). Hydraulics of Surface Irrigation, pp 449-498. In: *Design and Operation of Farm Irrigation Systems*, Edited by Jensen, ASAE

Brouwer, C.; Prins, K.; Kay, M. & Heibloem, M. (1988). *Irrigation Water Management: Irrigation Methods*, Training Manual No. 5, Food and Agriculture Organization of the United Nations, Available from www.fao.org/docrep/S8684E/S8684E00.htm (June 2011)

Burt, C.; Clemens, A. J.; Strelkoff, T. S.; Solomon, K. H.; Bliesner, R. D.; Hardy, L.A.; Howell, T.A. & Eseinhauer, D. E. (1997). Irrigation Performance Measures: Efficiency and Uniformity, Journal of Irrigation and Drainage Engineering, Vol. 123 (6), pp. 423-442 ASCE

EPA (2000). Environmental Response Team, In: Standard Operating Procedures, Soil Samples, Available from http://yosemite.epa.gov/sab/sabproduct.nsf/

Holzapfel, E. A.; Leiva, C.; Mariño, M. A.; Paredes, J.; Arumí, J. L. & Billib, M. (2010). Furrow Irrigation Management and Design Criteria Using Efficiency Parameters and Simulation Models, *Chilean Journal of Agricultural Research*, Vol. 70, Num. 2, pp. 287-296

Howell, A. T. (2002). *Irrigation Efficiency*, pp. 736-741, Marcel Dekker Inc. 270, Madison Avenue. New York, USA

Hsiao, T. C.; Steduto, P. & Fereres, E. (2007). A systematic and quantitative approach to improve water use efficiency in agriculture, *Irrigation Science*, Vol. 25, pp. 209-231, DOI 10.1007/s00271-007-0063-2

Humpherys, A. S. (1978). Improving Farm Irrigation Systems by Automation. *Proceedings of International Commitment on Irrigation and Drainage, 10th Cong.*, pp. 3590-3598, Athens, Greece

Khatri, K. L. (2007). *Toward Real-Time Control of Surface Irrigation*, Doctoral Dissertation, University of Southern Queensland, Toowoomba, Queensland, Australia

Luna, F. M. & Gutiérrez, S. R. (1997). *Informe Técnico*, INIFAP, Zacatecas.

Mojarro, D. F.; González, T. J.; Gutiérrez, N. J. A.; Toledo, B. A. & Araiza, E. J. A. (2004). *Software PIREZ (Proyecto Integral de Riego para el Estado de Zacatecas)*, Universidad Autónoma de Zacatecas, Mexico

Montiel-Gutiérrez, M. A. (2003). Estudio sobre la Eficiencia de Aplicación en el Distrito de Riego 038 "Río Mayo", Sonora, *XII Congreso Nacional de Irrigación*, pp. 27-38, Zacatecas, Mexico

Ojeda, B. W.; Hernández, B. L. & Sánchez, C. I. (2007). Requerimientos de Riego de los Cultivos, In: *Manual para Diseño de Zonas de Riego Pequeñas*, pp. 1-15, ISBN 978-968-9513-04-9, Comisión Nacional del Agua & Instituto Mexicano de Tecnología del Agua, México

Ortiz, V. J. G. (2005). *Evaluación en Campo del Método de Riego con Incremento de Gasto*, Master Degree Dissertation, Universidad Autónoma de Zacateca, Mexico

Playán, E. (2008). *Design, Operation, Maintenance and Performance Evaluation of Surface Irrigation Methods*, Land and Water Resources Management: Irrigated Agriculture, Istituto Agronomico Mediterraneo-CIHEAM, Italy

Rendón, P. L.; Fuentes, R. C. & Magaña, S. G. (1995). Diseño Simplificado del Riego por Gravedad, *VI Congreso Nacional de Irrigación*, pp. 385-391, Chapingo, México

Rendón, P. L.; Fuentes, R. C. & Magaña, S. G. (2007). Diseño del Riego por Gravedad, In: *Manual para Diseño de Zonas de Riego Pequeñas*, p. 75-86, ISBN 978-968-9513-04-9, Comisión Nacional del Agua & Instituto Mexicano de Tecnología del Agua, México

Rosano-Méndez, L.; Rendón, P. L.; Pacheco, H. P.; Etchevers, B. J. D.; Chávez, M. J. & Vaquera, H. H. (2001). Calibración de un Modelo Hidrológico Aplicado en el Riego Tecnificado por Gravedad, *Agrociencia* Vol. 35, pp. 577-588, ISSN 1405-3195

Sió, J.; Gázquez, A.; Perpinyà, M.; Peña J. C. & Virgili J.M. (2002). El uso eficiente del agua en el campo en Catalunya. El PACREG, un software para mejorar la gestión del agua del riego, El Agua y el Clima, Eds. J.A. Guijarro, M. Grimalt, M. Laita y S. Alonso, *III Congreso de la Asociación Española de Climatología*, Serie A, Num. 3, pp. 567-576. Palma de Mallorca, Spain

USDA (1956). *Methods for Evaluating Irrigation Systems*. Agriculture Handbook No.82, Soil Conservation Service, USA

Vázquez, F. E. (2001). Diseño del Riego con Incremento de Gasto en Surcos Cerrados, *Ingeniería del Agua*, Vol. 8, No. 3, pp. 339- 349, ISSN 1134-2196

Vázquez, F. E.; López, T. P.; Morales, S. E. & Chagoya, A. B. (2003). Primera Evaluación en Campo del Riego con Incremento de Gasto, *XII Congreso Nacional de Irrigación*, pp. 39-48, Zacatecas, México

Vázquez, F.E. (1996). *Formulación de un Criterio para Incrementar la Eficiencia de Riego por Surcos, Incluyendo el Método de Corte Posterior*, Doctoral Dissertation, DEPFI, UNAM.

Walker, W. R. & Skogerboe, G. V. (1987). *Surface Irrigation Theory and Practice*, Prentice-Hall, ISBN 0-13-877929-5, Englewood Cliffs, NJ, USA

Walker, W. R. (1989). *Guidelines for Designing and Evaluating Surface Irrigation Systems*, ISBN 92-5-102879-6 Irrigation and Drainage Paper 45, Food and Agriculture Organization of the United Nations, www.fao.org/docrep/T0231E/t0231e00.htm#Contents (June 2011)

3

Pumice for Efficient Water Use in Greenhouse Tomato Production

Miguel Angel Segura-Castruita[1], Luime Martínez-Corral[2],
Pablo Yescas-Coronado[1], Jorge A. Orozco-Vidal[1]
and Enrique Martínez-Rubín de Celis[1]
[1]Instituto Tecnológico de Torreón,
[2]Instituto Tecnológico Superior de Lerdo,
México

1. Introduction

The worldwide water demand is increasing (Yokwe, 2009). Although the type, magnitude and scope of the problems for water can vary from country to country (Biswas, 2010), the sector considered as the biggest user of water is agricultural (Hamdy et al., 2003). Therefore, water conservation through the efficient use of this resource in agriculture, is one of the most important firsts in the world (Díaz et al., 2004), especially in arid and semiarid regions (Forouzani & Karami, 2011). Different alternatives have been proposed to achieve sustainable water use in agriculture (Abad & Noguera, 2000; Díaz et al., 2004; Fuentes & García, 1999).

The food production in greenhouses is one of the alternatives to ensure efficient use of water (Abad & Noguera, 2000; Cadahia, 2000; Tahi et al., 2007). Greenhouses are structures that have a plastic cover, easy climate control systems, irrigation equipment and automated fertilization, elements that are used to increase productivity of horticultural and ornamentals crops (Cadahia, 2000; Cantliffe et al., 2003). In this production system, soil is often not used; instead other materials or mixtures of materials, known as substrate are used (Abad et al., 1996).

Substrates are mineral or organic materials, used for an anchor for plants and as a container of water they need (Dalton et al., 2006). To this end, substrates must have high storage capacity of water and keep it available to plants (Bender, 2008). However, these materials are often not native to the region where the activities of greenhouse production are taking place, which increases their cost (Cadahia, 2000). Therefore, natural materials and wastes are in a given region, have an important role in agricultural activities (Yaalon & Arnold, 2000) and in the preparation of substrates.

One of the naturally occurring mineral substrates, readily available and inexpensive is the sand, the material found in all environments (Moinereau et al., 1987). However, its moisture retention capacity is low, which implies the constant application of water to keep the plants grown on this substrate. Some industrial wastes such as rice husks, coconut fiber, coffee husks, rockwool, phenolic foams (Calderon & Cevallos, 2003), compost (López & Salinas, 2004) and pumice (Segura et al., 2008), also used as substrates.

Pome particles of 2-5 mm in diameter in natural conditions are responsible of the higher moisture holding capacity of sandy soils pomaceous origin (Segura, 2003), because they have a water storage capacity of 68%, of which 80% is readily available to plants (Segura et al., 2005). It is also suggested, that pumice can be used as promoter of retention of moisture in the soil or substrates for greenhouse, increasing efficiency in water use at low cost. In this regard, studies conducted without crop in sandy substrates of sedimentary origin with 30% of industrial pumice waste (2.38-3.35 mm in diameter) found increased water-holding capacity of the substrate 44%, of which, 56% was available (Segura et al., 2008). In this sense, the materials pomaceous discarded by blue jeans factories after the fabric softening and fuzz, become important in arid and semiarid regions of the world, where problems with water scarcity are found (Forouzani & Karami, 2011) and produces greenhouse vegetables to make an efficient water use. However, the behavior of moisture over time on this substrate with a horticultural crop has been rarely reported.

The tomato crop (*Lycopersicon esculentum* Mill) is one of the main vegetables grown throughout the world (Al-Omran et al., 2010; Bender, 2008), both in open ground and in greenhouse systems (Quintero, 2006). Water management in intensive production of tomato in greenhouse is of vital importance for this crop. Water performs a number of basic functions in the life of tomato plants, constituting 95% of its fresh weight (Castilla, 2005). In addition, the water needs of this plant are presented in three critical periods: in the emergence of the seedling, early flowering and during filling of fruit (González & Hernández, 2000).

In this context, it is likely that the amount of water used for development and production of tomato, decreases when using a sand-pumice substrate under greenhouse conditions; i.e. the growth and fruiting of tomato plants are not affected when applied different frequencies of irrigation due to capacity of moisture retention of sand-pumice substrate.

The objectives of this study were to evaluate the behaviour of moisture in a sandy-pumice substrate over time; assess the development of tomato plants until fruiting in this substrate in greenhouse and to establish the frequency of irrigation needed without affecting the development of plants in this substrate.

2. Materials and methods

2.1 Study area

The research was conducted in a greenhouse of Instituto Tecnológico de Torreón, located in the Región Lagunera, in Torreón, Coahuila México. This region is located between 25° 36′ 45″ and 25° 36′ 47″ north latitude and between 103° 22′ 19″ and 103° 22′ 21″ west longitude, at 1159 meters above sea level (Fig. 1). Climate is a dry desert with summer rainfall and cool winter [Bw (h′) hw (e)] (García, 1988). Total annual precipitation is 250 mm and evaporation is 2 400 mm. The average annual temperature is 21 °C, with annual maximum temperature of 29.5 °C and minimum of 14.8 °C; the warmest month is June (35.3 °C) and January is the coldest month with 6.6 °C (CNA, 2011).

2.2 Methodology

The work was divided in three stages: 1) Selection of experimental material, 2) Greenhouse work and 3) Laboratory work.

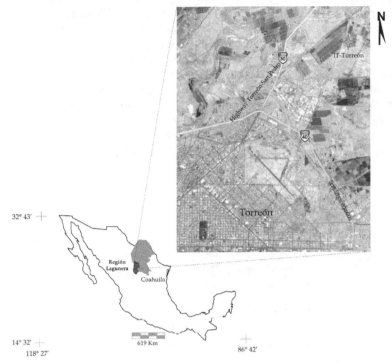

Fig. 1. Localization of study area

2.2.1 Selection of the experimental material

The substrate used was a mixture of sand with pumice. The sand was obtained by sieving coarse sediments of the Nazas river through two sieves (2 mm and 0.5 mm in diameter) to remove particles of the size of gravels, silts and clays, as established in the laboratory manual of Soil Survey Laboratory (SSI, 1996). The mineralogical composition of the sand was obtained by powder diffraction of X-ray (Melgarejo et al., 2010). The sand has quartz, pyroxene, biotite, mica, and calcium and sodium feldspar. The sand retains water easily available, but quickly lost its due to its low moisture retention capacity (Calderon & Cevallo, 2003; Porta et al., 1999). Carbonates were removed from sand (Resh, 1989). Pumice particles were obtained from wastes of a maquiladora company in the region. Pumice wastes were crushed and sieved to bring them to a diameter between 2.38 mm and 3.35 mm, size of the particles that retains a greater amount of water (60% in their pores) (Segura, 2003). Subsequently, particles of desired size were put through a wash, cool water first and then with hot water in order to eliminate industrial wastes (Sandoval & Brisuela, 2002); finally they left to dry in the shade at room temperature. Tomato plants (*Lycopersicon esculentum* Mill) cv. Río Grande, were used in this experiment. This variety has 90% germination, is of intermediate and semi-early cycle (Paez et al., 2000).

2.2.2 Greenhouse work

The sand-pumice substrate was elaborated according to methodology used by Segura et al. (2008), with the proportions of 30% pumice and 70% sand, based on volume. This substrate has a field capacity of 15.7% and a permanent wilting point of 4.9, with available moisture of 10.8% (Segura et al., 2008). The substrate was placed in black plastic pots (circular area of 283.53 cm²) with a capacity of six litres and previously plugged holes. The weight of dry substrate in the pot was recorded in the balance, resulting in an average weight of five kilograms per pot. Also, before transplantation of tomato seedlings, the substrate in the pots was saturated with water for 48 hrs. When time went on, plugged holes of the pots were uncovered in order to drain excess water until the drip rate was one drop every ten seconds (Preciado et al., 2002) ensuring that the substrate was on field capacity (Segura et al., 2008). At this point, the weight of the wet pot was recorded, so that subtracting to this data the weight of the pot with dry substrate it is obtained the initial water weight or initial moisture content.

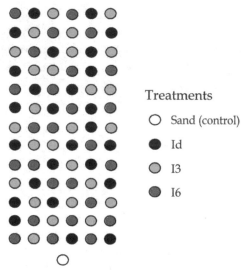

Treatments

○ Sand (control)

● Id

○ I3

● I6

Fig. 2. Distribution of the treatment in the experiment

Tomato seeds were placed in a container whose cavities had substrate (peat moss) wet. Light irrigations with modified nutritious solution of Steiner (1961) whose components are N, P, K, Ca, Mg, Fe, Cu, Zn, Mn and B, with 550, 27, 514, 634, 122, 10, 0.5, 0.45, 1.23 and 0.29 mg·L⁻¹, respectively (Magdaleno-Villar et al., 2006), was applied every three days until the seed germinated. The transplant was performed when the seedlings were 20 cm in height, measured from the cotyledons. The plants before being transplanted were weighed, with the aim of adding weights to the initial moisture and monitor the behaviour of moisture later.

The plants in the substrate were put through three different irrigation frequencies: daily, every three days and six days (I_d, I_3 and I_6, respectively), which for this experiment corresponded to the treatments. Each frequency had 26 repeats and a control with only sand and frequency of irrigation daily (control), so that the work consisted of 79 experimental units. The experimental design was completely random (Fig. 2).

The moisture content of each pot was recorded daily by the gravimetric method at twelve hours for eighty days. The lost water was replenished with nutritious solution according to the corresponding treatment; for example, the difference in moisture in the pot of I_d treatment was recovered at the time of the weighing. When the treatment was I_3, the moisture content was recorded only, and the solution was added until the fullness of the set time and so on.

At the same time, destructive sampling of plants in each treatment were carried out at 28, 40, 52 and 60 days after transplanting (dat) according to proposed by Segura et al. (2011). Five plants of each treatment were randomly selected in each set date. Plants were separated from the substrate to obtain the data of plant height, root length, number of flowers and number and weight of fruit. The average fresh weight of plants per treatment at each sampling was added to the rest of the corresponded pots, in order to have a total weight of each of them and continue with their gravimetric weight every day. Finally, at 80 dat six plants were evaluated leaving a potted plant to another process in the laboratory. The results were evaluated by analysis of variance and a Tukey's test at $p \leq 0.05$.

2.2.3 Laboratory work

At the end of the greenhouse stage, the substrate of the last pot with plant of each treatment was analyzed micromorphologically in order to observe its porosities and spatial distribution of the particles of pumice and sand. To achieve this, the pots were taken to the laboratory without disturbing the physical conditions of the substrate, i.e. without moving the substrate; only the aerial part of the plat was cut at the base (two centimetres above the surface of the substrate).

The four pots with substrate and plant roots were air-dried and in the shade. Once dried, they were impregnated with polyester resin and styrene monomer, at a ratio of 7:3, mixed with potassium fluoride. Subsequently, they were left in gelation in the shade for 20 days.

When they were hardened, it was proceed to cut them with a diamond blade and polished with different abrasives to a thickness of 30 µm, to obtain thin sections of 6×7.5 cm. The thin sections were analyzed with an Olympus petrographic microscope, with magnifications of 2.5× to 100×. The description of the thin sections and microconstituents was based on the manual developed by Bullock et al. (1985). The description of the porosity was performed on digital images obtained with a CCD Olympus camera of 4.1 Megapixel, getting close up of rectangular shape array of 86×64 mm (5504 mm^2) with a spatial resolution of 31 µm per pixel. The analysis of the image was performed with an analyzer Image Pro Plus v4.5 (Media Cybernetic, Maryland, U.S.A.).

In the other hand, the average leaf area for each treatment was obtained by cutting eight leaves per plant sampled; on the leaves the maximum length was measured from the base of the petiole up to the tip of the central leaflet and the maximum width of the leaves was measured perpendicular to maximum length (Astegiano et al., 2001). In addition, the leaves were photocopied to obtain leaf area by using a LICOR (LI-3000) leaf area meter.

Also, the efficiency of water use (EWU) of tomato plants was obtained by relating the gross weight of the fresh fruits in kilograms, until the last day of the experiment, with the total amount of water (in m^{-3}) used until then.

3. Results and discussion

3.1 Behaviour of moisture in the substrate

The amount of water used in each treatment was different (see Table 1). The water consumption presented when using sand-pumice substrate at different irrigation frequencies, was lower than the substrate of sand (98.82 L plant[-1]). The above was due to the effect that the pumice particles had with sand and water. When moisture comes into contact with mineral particles create a potential difference matrix (Hillel, 1982; Miller & Gardner, 1962). In this sense, the sand-pumice substrate is a system composed of two subsystems, where the sandy substrate has greater potential than the pumice, which causes the water retained in the sand evaporates first and then the one in the pumice; i.e. the water in the pores of the sand, product of arrangement of its particles, evaporates first and subsequently the one found in the pores of the pumice (Segura et al., 2008).

Sample date (days)	Water applied (mm)							
	Sand substrate		Sand-pumice substrate					
			I_d		I_3		I_6	
	Total	Daily average water	Total	Daily average water	Total	Daily average water	Total	Daily average water
0	0	0	0	0	0	0	0	0
28	6.52	0.23	2.71	0.09	2.03	0.07	1.45	0.05
42	4.01	0.28	3.21	0.22	2.98	0.15	2.13	0.10
52	5.60	0.56	2.93	0.28	2.51	0.25	1.88	0.18
60	4.91	0.61	2.63	0.32	2.31	0.28	2.11	0.26
80	13.80	0.69	8.81	0.42	7.69	0.38	6.51	0.32
Total	34.85		20.29		17.55		14.10	

Table 1.Water applied to plants tomatoes in sand and sand-pumice substrates and different irrigation frequencies

Pomaceous material found in sand acts as little tightly sponges (Daniels & Hammer, 1992) with a high content of pores (60%) of which, 15% are driving pores and 45% are storing pores (Segura et al., 2003) or mesopores (30-70 μm in diameter) and micropores (<30 μm in diameter), respectively (Sumner, 2000).

Analyzing micromorphologically sand-pumice substrate, it was observed an apedal structure (without aggregates). In this context, porosity is defined as of compound packaging (Bullock et al. 1985; Stoops, 1993), because of presence of pomaceous particles with interconnected pores, which causes the existence of a greater number of pores of storage in sand substrate (Segura et al., 2003; Sumner, 2000). Figures 3 and 4 are examples of the above.

In Fig. 3 it shows the exposure of a pumice particle in 20×, where it can be seen a rough surface with pores of different diameter and a magnification at 100× of a thin section where it appears the porosity of pumice, place where water is stored. In Fig. 4, three photomicrographs (in greyscale, in ultraviolet light and binary format) show spatial distribution of particles of pumice and sand, in this case the clearer parts on the picture 4a and 4b belongs to porosity space; see as how the pore space is increased where the pumice is.

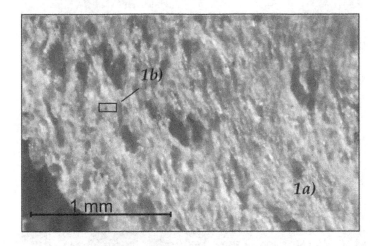

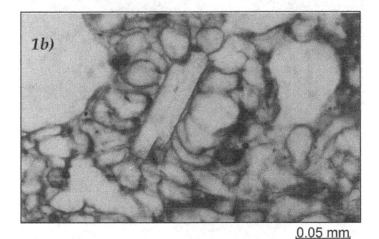

Fig. 3. Photomicrographs of a pumice particle: *1a*) pumice particle with incident light, *1b*) pumice porous in thin section with plane polarized light

Even when different treatments had the same substrate, in treatment I_d (sand-pumice substrate) was missed a greater amount of water (20.29 mm·plant⁻¹) than treatments in frequencies I_3 and I_6 (17.55 y 14.10 mm·plant⁻¹, respectively). Segura et al. (2008) indicate that the moisture in the sand-pumice substrate is lost at a rate of 2.46% per day. In our case, each treatment had higher ratios of percentages of loss than those reported in the literature. Considering that initial moisture was of 0.74 mm, I_d, I_3 and I_6 treatments had 11.43%, 10.00% and 9.06% of loss of water daily, respectively.

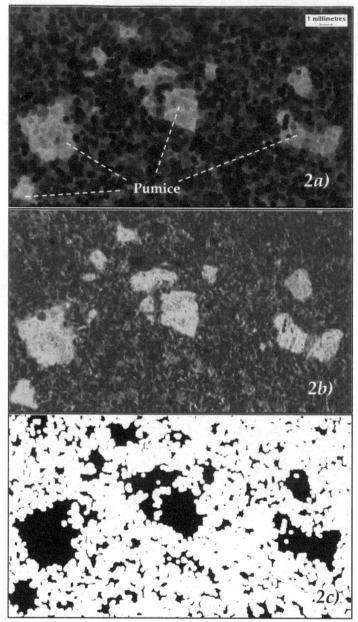

Fig. 4. Photomicrographs of sand-pumice substrate: 2*a*) greyscale; 2*b*) ultraviolet light and 2*c*) binary format. The pumice particles increase the porous space of sand

Although, it must be considered that the effect of plant factor used in this work, since in those reported in the literature all that was evaluated was the substrate. For this reason, the treatments had different loss rates (see Table 2), tending to reduce the average moisture as time went on in days. This behaviour was mainly due to the different times when irrigation was applied, where it could be observed the effect of pumice on the moisture retention capacity of sand and therefore, the amount of water to replenish (Fig. 5). In our case, the models of lineal regression between the moisture of substrate and I_d, I_3 and I_6 treatments (MI_d, MI_3 and MI_6, respectively) and time (t) show the effect of timing of irrigation in the substrate.

Treatment	$M = \beta t + c$	R^2	p^*
Sand	$Ms = -0.0625\ t + 0.2461$	0.7222	0.058
I_d	$M_{Id} = -0.0847\ t + 0.7212$	0.9749	0.048
I_3	$M_{I3} = -0.0741\ t + 0.7298$	0.9883	0.022
I_6	$M_{I6} = -0.0671\ t + 0.7521$	0.9949	0.035

*p: Rejection probability of the variance analysis regression.

Table 2. Simple lineal regression model between the H to each treatment and t

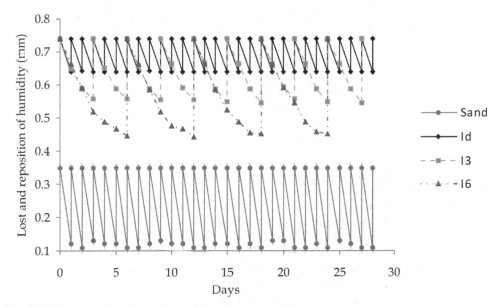

Fig. 5. Behaviour of moisture in each treatment over time

These models are significant at $p \leq 0.05$ level. The rejection probability value ($p = 0.022$) of the analysis of variance of the regression between H_{I3} and t was the most significant, with a negative trend and a $R^2 = 0.9883$. However, these results are averages per day and differ from those reported by other authors (Savvas et al., 2006; Tzortzakis & Economakis, 2007). When analyzing the average amount of water applied by irrigation time over time, in I_6 treatment were applied larger volumes of water divided into a few times of application (see Table 3).

Treatment	Sample date (days)	Number of irrigation for each period	Average of water applied by irrigation (mm)
I_d	28	28	0.10
	42	14	0.22
	52	10	0.29
	60	8	0.32
	80	20	0.44
I_3	28	9	0.22
	42	5	0.59
	52	3	0.83
	60	2	1.15
	80	7	1.09
I_6	28	5	0.29
	42	2	1.06
	52	2	0.94
	60	1	2.11
	80	3	2.17

Table 3.Average amount of water applied to each irrigation time and the number of irrigation between sampling and sampling

This situation is contrasting with those in I_d, where volumes were lower, even though the total of volume in this treatment was higher than in other frequencies. As an example figure 3 is shown, where it can be seen that the lost of moisture in the treatments I_3 and I_6 were accumulating exponentially, making this the action of pumice, recovering the lack of water at time is was supposed to. In the I_d treatment the water was resting at the time of its weighted, so it was not observed the moisture loss exponentially as in other treatments. Segura et al. (2011) indicate that in daily irrigation is not given an opportunity to the functioning of the sand-pumice system, since moisture is replenished day after day; whereas when irrigation is every three days, the moisture is lost first in the sand remaining in the pumice. This situation goes so far when irrigation is every six days, since water is lost in the sand and in the pumice. This can result in problems in the development of plants (Tahi et al., 2007).

All the above indicates that the sand-pumice substrate has the ability to store water and not lose easily, but what happened to the tomato plants that were in the substrate? This topic will be addressed below.

3.2 Development of tomato plants
3.2.1 Water consumption
The different tomato plants during its development and fruiting, consumed different amounts of water at each sampling date (see Table 1). In the I_d treatment the amount of average water per pot with tomato plant in the first sampling (28 dat) was of 0.09 mm·day^{-1}·plant^{-1} and 0.42

mm·day^{-1}·plant^{-1} from the 80 dat; these results have little difference to those reported in the literature. Flores et al. (2007) indicate that the tomato plant in the early days of development requires 0.200 L (0.07 mm) and during the period of greatest demand up to 1.500 L (0.52 mm). This contrasts with findings in the treatments I_3 and I_6. For example, the I_3 treatment in the period of greatest demand (from 60 to 80 dat) plants consumed 0.38 mm·day^{-1}·plant^{-1}, i.e. there was a water savings of 10% compared with I_d treatment. This can be explained by the moisture retention capacity of the substrate, as mentioned earlier.

The I_6 treatment had a lower increase due to water requirements by tomato plants as time passed. In this regard, when water stress is generated in tomato crop, the plant reacts by closing its stomata to avoid perspiration (Al-Omran et al., 2010; Asgharipour & Armin, 2010; Bender, 2008). However, if water stress is prolonged, the plant is able to transpire accumulating solutes and reducing the size of their cells to decrease the water potential; when this happens, the plant opens its stomata partially to continue with its vital functions (Reddy et al., 2005). In our case, the results in this treatment at 60 dat was of 73 cm, result similar to that report (75 cm) by Macua et al. (2003) in plants subjected to water stress.

Treatment	Sample date				
	28	42	52	60	80
	Leaf area (cm^2)				
Sand	250.23 a	1340.50 a	1453.61	1540.22	1534.25
I_d	252.03 a*	1345.08 a	1458.82 a	1542.13 a	1548.44 a
I_3	255.98 a	1221.11 a	1354.65 b	1469.42 b	1559.57 a
I_6	93.52 b	586.81 b	710.85 c	739.56 c	801.71 b
LSD	56.45	289.37	77.35	22.15	31.13
	Root lenght (cm)				
Sand	20 a	35 a	47 b	54 a	60 a
I_d	22 a	38 a	50 a	57 a	65 a
I_3	20 a	39 a	48 b	55 a	66 a
I_6	17 b	26 b	28 c	32 a	39 b
LSD	2.88	4.88	1.90	3.21	6.50

*Different letters in the same column indicate significant difference, Tukey's test ($p \le 0.05$). LSD: Lower significant difference

Table 4. Results of leaf area and root length of tomato plants in sand-pumice substrates at different sampling dates

In the case of leaf area (La) the trend of results was similar to plant height. The I_d treatment had the highest La at the end of the experiment (1548.44 cm^2); however, it did not present significant differences ($p \le 0.05$) compared to I_3 treatment (see Table 4). The I_6 treatment presented the lowest La in the experiment. Orozco et al. (2008) indicated that the leaf area is a physical indicator that determines the magnitude of the photosynthetic machinery, used to meet the demand of photosynthate by the growing organs of the crop and for this; the plant requires the presence of water to carry out the process of photosynthesis. The absence of significant differences between the treatments of daily irrigation and every third day at the end of the experiment is an indication that despite not adding water daily in treatment I_3, tomato plants were not subjected to a water stress.

When tomato plant was not exposed to prolonged water stress, it can perform its physiological processes without any problems (León et al., 2005; Páez et al., 2000; Tahi et al., 2007). Instead, the I_6 treatment by presenting a lower La, reflects the water stress to which it was submitted. In general, the lack of water results in a poor development of leaf area since water reservoir that a plant has is used to stay alive and thus decrease its physiological processes (Macías et al., 2010; Sirvansa, 2000; Tahi et al., 2007).

Another physiological variable evaluated was the root. The length of this organ in the experiment was significant ($p \leq 0.05$). At first differences among the three treatments were presented, but after the 52 dat the I_d and the I_3 treatments had very similar root lengths unlike the I_6 treatment (see Table 4). Results closely related to the presence of water in the substrate, as explained above.

Fig. 6. Tomato plant roots in sand-pumice substrate

One of the interesting aspects that were observed was the presence of roots in the particles of pumice (see Fig. 6). Several authors have reported that plant roots may have access to the water that is found inside the pores of pumice (De León et al., 2007; Savvas et al., 2006). Event that occurs because the water in the pores of particles of pumice is held at a tension less than 0.0024 kPa, so the water is readily available (Segura et al., 2008).

3.2.2 Flowers and fruits

The flowers of tomato in this experiment appeared after the 28 dat (see Table 5). The number of flowers between I_d and I_3 treatments had not significant differences ($p \leq 0.05$) until eighty days. However, it can be observed that the amount decreases in accordance with the passing of time, this condition occurs because it began with fruiting. León et al. (2005) state that moderate water stress does not affect physiological process in tomato plants as flowering and fruiting (Tahi et al., 2007). The I_6 treatment introduced its first flowers after the sixty days. The absence of flowering plants in this treatment was due to water stress to which they were subjected. The tomato plants are sensitive to water stress and high temperatures, since they affect flowering and reduce fruit production (Shubang, 2002; Sirvansa, 2000).

Treatment	Sample date			
	42	52	60	80
	Flowers			
Sand	13 a	17 a	23 a	16 a
I_d	12 a*	16 a	22 a	17 a
I_3	13 a	18 a	20 b	16 a
I_6	-	-	-	9 b
LSD	2.0	5.1	2.0	3.3
	Fruits			
Sand	-	-	8 a	23 a
I_d	-	-	8 a	22 a
I_3	-	-	5 b	21 a
I_6	-	-	-	-
LSD	-	-	2.0	2.0

*Different letters in the same column indicate significant difference, Tuckey's test ($p \leq 0.05$).

LSD: Lower significant difference

Table 5. Results count flowers and fruits of tomato plants in sand-pumice substrates at different sampling dates

The appearance of the fruits started to sixty days. Between the I_d and I_3 treatments there were no significant difference at 80 dat. On average there were 22 and 21 fruits per plant unharvested, respectively (Table 5). The fruits harvested totalled 15 for the I_d treatment and 14 for the I_3 treatment per plant, with an average weight per fruit of 140 and 135 g, respectively.

3.3 Efficient water use (EWU)
The EWU of tomato plants showed significant differences between the control and the other treatments under study. The EWU of the plants in the sand substrate was 36.32 g·m^{-3} (Table 6), result in treatment less than in I_d and I_3 treatments (36.84 and 38.57 g·m^{-3}, respectively). These results are greater than (12.1 g·m^{-3}) reported by other authors (Al-Omran et al., 2010; Tahi et al., 2007). Even though the experiment lasted only eighty days, the results showed

the usefulness of sand-pumice substrate; also demonstrate the utility of pumice for this purpose.

Substrate	Treatment	Tomato plant yield to 80 dat (g)	Total water applied (m^{-3})	EWU (g·m^{-3})
Sand	Daily	3.560	0.098	36.32 a*
Sand-pumice	I$_d$	2.100	0.057	36.84 a
	I$_3$	1.890	0.049	38.57 a
	LSD			3.36

*Different letters in the same column indicate significant difference, Tukey's test ($p \leq 0.05$).

LSD: Lower significant difference.

Table 6. Efficient water use of tomato plants in sand and sand-pumice substrates

4. Conclusions

The behaviour of moisture in the sand-pumice substrate with tomato plants is consumed at a rate of 10 to 11% per day with respect to its ability to retain moisture. This allows to space the time of application of water every three days to tomato plants on this substrate without its development is affected, so the flowering and fruiting stages are carried out without putting plants to a water stress. The use of pumice particles as an improver of moisture holding capacity of sandy substrate helps plants to make an efficient use of water in greenhouse. However, more research related to nutritional quality of fruits is needed to ensure the obtaining of a quality product making an efficient use of water.

5. References

Abad B., M. & Noguera, P. (2000). Sustratos para el cultivo sin suelo y fertirrigación. In: *Fertirrigación. Cultivos hortícolas y ornamentales.* 2ª. Ed. C. Cadahia L., (ed.), 215-216, Ediciones Mundi-Prensa, ISBN 978-484-7624-75-8, D.F., México.

Abad, M.; Noguera, P. & Noguera, V. (1996). Turbas para semilleros. *In: II Jornadas sobre semillas y semilleros hortícolas. Congresos y Jornadas, 35/96.* Juntas de Andalucía. Consejería de Agricultura y Pesca. pp. 52-53, ISBN 84-875664-40-2, Sevilla, España, Mayo 29-31, 1995.

Al-Omran, A.M.; Al-Harbi, A.R.; Wahb-Allah, M.A.; Nadeem, M. & Al-Eter, A. (2010). Impact of irrigation water quality, irrigation systems, irrigation rates and soil amendments on tomato production in sandy calcareous soil. *Turk Journal Agriculture,* Vol. 34: 59-73.

Asgharipour, M.R. & Armin, M. (2010). Growth and elemental accumulation of tomato seedlings grown in composted solid waste soil amended. *American-Eurasian Journal of Sustainable Agriculture,* Vol. 4: 94-101.

Astegiano, E.D.; Favaro, J.C. & Bouzo, C.A. (2001). Estimación del área foliar en distintos cultivares de tomate (*Lycopersicon esculentum Mill.*) utilizando medidas foliares. *Invest. Agr.: Prod. Prot. Veg.*, Vol. 16: 249-256.

Bender Ö., D. (2008). Growth and transpiration of tomato seedlings grown in Hazelnut Husk compost under water-deficit stress. *Compost Science & Utilization*, Vol. 16: 125-131.

Biswas, A.K. (2010). Water for thirsty urban World. *Brown Journal of World Affairs*, Vol. 17: 147-166.

Bullock, P.; Federoff, N.; Jongerius, A.; Stoops, G. & Tursina, T. (1985). *Handbook for Soil Thin Section Description*. Wine Research Publications, ISBN 0-905-18409-2, Albrighton Wolverhampton, UK.

Cadahia L., C. (2005). Fertirrigación. Cultivos hortícolas, frutícolas y ornamentales. 3ª. Ed. pp. 123-124, Ediciones Mundi-Prensa, ISBN 978-484-7624-75-8 D.F., México.

Calderon S., F. & Cevallos, F. (2003). Los Sustratos. In: *Memorias del Primer Curso de Hidroponía para la Floricultura*, Ene 15, 2011,
http://www.drcalderonlabs.com/Publicaciones/Los_Sustratos.htm

Cantliffe, D.J.; Funes, J.; Jovicich, E.; Parajpe, A.; Rodríguez, J. & Shaw, N. (2003). Media and containers for greenhouse soilless grown cucumbers, melons, peppers and strawberries. *Acta Hort.*, Vol. 64: 199-203.

Castilla, N. (2005). Invernaderos de plástico, tecnología y manejo. pp. 245-246, Ediciones Mundi-Prensa, ISBN 84-8476-221-1, Madrid, España.

Dalton, G.S., McMaster, J.S. & McMaster, L.C. (2006). Developing and implementing a biodiversity strategy for processing tomato farms. *Acta Hort.*, Vol. 724: 207-213.

Daniels, R.B. & Hammer, R.D. (1992). *Soil Geomorphology*. Wiley, ISBN 0-471-5113-6, Nueva York, USA.

De León G., F.; Gutiérrez C., M.C.; Gónzalez CH., M.C. & Castillo J., H. (2007). Root aggregation in a pumiceous sandy soil. *Geoderma*, Vol. 142: 308-317.

Díaz O., A.; Escalante E., J.A.; Trinidad S., A.; Sánchez G., P.; Mapes S., C. & Martínez M., D. (2004). Rendimiento, eficiencia agronómica del nitrógeno y eficiencia en el uso del agua en amaranto en función del manejo del cultivo. *Terra Latinoamericana*, Vol. 22: 109-116.

Flores, J.; Ojeda B., W.; López, I.; Rojano, A. & Salazar, I. (2007). Requerimientos de riego para tomate en invernadero. *Terra Latinoamericana*, Vol. 25: 127-134.

Forouzani, M. & Kapani, E. (2001). Agricultural water poverty index and sustainability. *Agronomy for Sustainable Development*, Vol. 31: 415-432.

Fuentes Y., J.L. & García L., G. (1999). *Técnicas de Riego. Sistemas de Riego en la Agricultura*. Mundi-Prensa, ISBN 978-968-7462-17-2, D.F., México.

García, E. (1988). *Modificaciones al sistema de clasificación climática de Köppen*. UNAM-Instituto de Geografía, D.F., México.

González M., A. & Hernández L., B.A. (2000). Estimación de las necesidades hídricas del tomate. *Terra*, Vol. 18: 45-50.

Hamdy A., Ragab R., Scarascia-Mugnozza E. (2003) Coping with water scarcity: water saving and increasing water productivity. *Irrigation Drainage*, Vol. 52: 3–20.

Hillel, D. (1982). *Introduction to soil physics*. Academic Prees. ISBN 978-012-3485-20-5, Orlando, Fl., USA.

León, M.; Cun, R.; CHaterlan, Y. & Rodríguez, R. (2005). Uso del agua en el cultivo del tomate protegido. Resultados obtenidos en Cuba. *Revistas Ciencias Técnicas Agropecuarias*, Vol. 14: 9-13.

López G., J. & Salinas A., J.A. (2004). *Efectos Ambientales del Sistema de Cultivo Forzado*. In: *Encuentro Medioambiental Almeriense*. Rivera M., J. (Ed), ISBN 0-848-2401-67-X, Universidad de Almería, Almería, España.

Macías D., R.; Grijalva C., R.L. & Robles C., F. (2010). Efecto de tres volúmenes de agua y la productividad y calidad de tomate bola (*Lycopersicon esculentum* Mill) bajo condiciones de invernadero. *Biotecnia*, Vol. 12: 11-19.

Macua, J.I.; Lahoz, I.; Arzoz, A. & Gamica, J. (2003). The influence of irrigation cut-off time on the yield and quality of processing tomatoes. *Acta Hort.*, Vol. 613: 151-153.

Miller, D.E. & Gardner, W.H. (1962). Water infiltration into stratified soil. *Proc. Soil Sci. Soc. Am.*, Vol. 26: 115-118.

Magdaleno V., J.J.; Peña-Lomeli, A.; Castro-Brindis, R.; Castillo-González, A.M.; Galvis-Spinola, A.; Ramírez-Pérez, F. & Hernández-Hernández, B. (2006). Efecto de soluciones nutritivas sobre el desarrollo de plántulas de tomate de cáscara (*Physalis ixocarpa* Brot.). *Revista Chapingo Serie Horticultura*, Vol. 12: 223-229.

Malgarejo, J.C.; Proenza, J.A.; Gali, S. & Llovet, X. (2010). Técnicas de caracterización mineral y su aplicación en exploración y explotación minera. *Boletín de la Sociedad Geológica Mexicana*, Vol. 62: 1-23.

Moinereau, J.; Hermann, J.P.; Favrot, C. & Riviere L.M. (1987). Les substrats-Inventaire, caractéristiques, ressources. En *Les Cultures Hors Sol*. 2e ed. Blanc D (Dir.), pp. 15-75. Institut National de la Recherche Agronomique, París, Francia.

Orozco V., J.A.; Palomo G, A.; Gutiérrez R., E.; Espinosa B., A. & Hernández H., V. (2008). Dosis de nitrógeno y su efecto en la producción y distribución de biomasa de algodón transgénico. *Terra-Latinoamericana*, Vol. 26: 29-35.

Páez, A.; Paz, V. & López, J.C. (2000). Crecimiento y respuestas fisiológicas de plantas de tomate cv. Río Grande en la época mayo-julio. Efecto del sombreado. *Rev. Fac. Agron. (LUZ)*, Vol. 17: 173-184.

Porta C., J.; López-Acevedo R., M. & Roquero L., C. (1999). *Edafología, para la Agricultura y el Medio Ambiente*. Mundi-Prensa, ISBN 978-847.1147-84-4, D.F., México.

Preciado R., P.; Baca C., G.; Tirado T., J.L.; Kahuasi, S.J.; Tijerina CH., L. & Martínez G., A. (2002). Nitrógeno y potasio en la producción de plántulas de melón. *Terra*, Vol. 20: 267-276.

Quintero, M.F.; González, C.A. & Florez R., V. J. (2006). Physical and hydraulic properties of four substrates used in the cut-flower industry in Colombia. *Acta Hort.*, Vol. 718: 499-506.

Reddy, A.R.; Chaitanya, K.V.; Jutur, P.P. & Granam, A. (2005). Photosynthesis and oxidative stress responses to water deficit in five different mulberry (Morus alba L.) cultivars. *Physiol. Mol. Biol. Plants*, Vol. 11: 291-298.

Resh, H. (2005). *Hyddroponic food production*. 5ª ed. Woodbridge, ISBN 0-931231-99-X, Sta. Barbara, CA, USA.

Sandoval V., M. & Brisuela A., P.B. (2002). Horticultura intensiva en invernaderos. *XXXI Congreso Nacional de Ciencia del Suelo*, pp. 283, Torreón, Coahuila, México. 13-17 Octubre 2002.

Savvas, D.; Passam, H.C.; Olympios, C.; Nasi, E.; Moustaka, E.; Mantzos, N. & Barouchas, P. (2006). Effects of ammonium nitrogen on lettuce grown on pumice in a close hydroponic system. *HortScience*, Vol. 41: 1667-1673.

Segura C., M.A. (2003). *Escalas de observación en los estudios de génesis de suelos: caso de los suelos de humedad residual*. Tesis de Doctorado. Colegio de Postgraduados. Montecillo, México.

Segura C., M.A.; Gutiérrez C., M.C.; Ortiz S., C.A. & Sánchez G., P. (2005). Régimen de humedad y clasificación de suelos pomáceos del Valle Puebla-Tlaxcala. *Terra Latinoamericana*, Vol. 23: 13-20.

Segura C., M.A.; Ortiz S., C.A.; Gutiérrez C., M.C. (2003). Localización de suelos de humedad residual con el uso de imágenes de satélite: Clasificación automática supervisada de la imagen. *Terra*, Vol. 21: 149- 156.

Segura C., M.A.; Preciado R., P.; González C., G.; Frías R., J.E.; García L., G.; Orozco V., J.A. & Enríquez S., M. (2008). Adición de material pomáceo a sustratos de arena para incrementar la capacidad de retención de humedad. *INTERCIENCIA*, Vol. 33: 923-928

Segura C., M.A.; Ramírez S., A.R.; García L., G.; Preciado R., P.; García H., J.I.; Yescas C., P.; Fortis H., M.; Orozco V., J.A. & Montemayor T., J.A. (2011). Desarrollo de plantas de tomate en un sustrato de arena-pómez con tres diferentes frecuencias de riego. *Revista Chapingo Serie Horticultura*, Vol. 17 (Especial 1): 25-31.

Shubang, N. (2002). Effect of water stress during flowering on macademia plants. *J. Southwest Agric. Univ.*, Vol. 24: 34-37.

Sirvansa, R. (2000). Tolerance to water stress in tomato cultivars. *Photosyntetica*, Vol. 38: 465-467.

SSI Soil Survey Investigation. (1996). *Soil Survey Laboratory Methods Manual*. Report No. 42 Version 3. U.S.D.A. N.R.C.S. and N.S.S.C. U.S. Government Printing Office, Washington, D.C., USA.

Steiner, A.A. (1961). A universal method for preparing nutrient solutions of a certain desired composition. *Plant Soil*, Vol. 15: 134-154.

Stoops, G. (1993). *Lectures of Micromorphology*. International Training Center. University of Ghent. Gent, Bélgica.

Sumner, M. (2000). *Handbook of Soil Science*. CRC, ISBN 084-9331-36-6, Nueva York, USA.

Tahi, H.; Wahbi, S.; Wakrim, R.; Agachich, B.; Serraj, R. & Centritto, M. (2007). Water relations, photosynthesis, growth and water-use efficiency in tomato plants subjected to partial rootzone drying and regulated deficit irrigation. *Plant Biosystems*, Vol. 141: 265 - 274.

Tzortzakis, N.G. & Economakis, C.D. (2007). Maintaining postharvest quality of the tomato fruit by employing methyl jasmonate and ethanol vapor treatment. *Journal of Food Quality*, Vol. 30: 567-580.

Yaalon, D.H. & Arnold, R.W. (2000). Attitudes toward soils and their societal relevance: then and now. *Soil Science*, Vol. 165: 5-12.

Yokwe, S. (2009). Water productivity in smallholder irrigation schemes in South Africa. *Agricultural Water Management*, Vol. 96: 1223-1228.

4

Urbanization, Water Quality Degradation and Irrigation for Agriculture in Nhue River Basin of Vietnam

Mai Van Trinh and Do Thanh Dinh

Institute for Agricultural Environment,
Vietnam

1. Introduction

The economic develope with a number of different agricultural, industrial productions, services and comercial activities, but at the same time river basins are highly populated. Because of that, environmental issues in these rivers are very important and many people care about waste, pollution and environmental protection. Pollution from animal waste, for example, Karr et al. (2003) used radio-active isotope ^{15}N to study the transportation and contribution of pollutants from high populated graizing lands to environmental quality of Neuse river basin and showed that ^{15}N can be used well in ditermining the sources of nitrate pollution. Gardi (2001) used Geographic Information System (GIS) and observed data on water discharge, sediment, plant protection chemical and nitrate conentration in the watershed near by Bologna. By simulating agricultural activities, the use of chemicals and fertilizers research found a residue of chemical within the watershed higher than European standard. Research also found that increasing the area of row planting crops increasing the chemical residue in the watershed water. Kraft and Stiles (2003) reported that at the tropical regions the light soil texture and shallow ground water table conditions cause high vulnerability of nitrate pollution. It is recommended that people in the region need to have a management and monitoring system for production activities and minimize the input factors to mitigate nitrate pollution. Studying the pathway and pollution of phosphorus at a watershed in the western USA, McDowell and Sharpley (2003) concluded that they need to apply integrated measures on management of manures, agricultural farming and developing natural filters and strengthen production management to mitigate environmental pollution. Using Nutrient Transport and Transformation (NTT) to simulate the transpotation of nitrogen at Muddy Brook watershed in north-west of Connecticut, USA, Nikolaidis et al. (1998) based on the research results to give a management guideline to choose suitable managements to improve water quality in the river and tream network. Modelling is a good trend to assess the current environment, validate the results of management and planning measures in the watershed, for example, mass of pollutants from Glumso lake, Denmark were also simulated (Zhang and Jorgensen, 2005) using Pamolare model; use of GIS and Agricultural Non-Point Pollution Source (AGNPS) to assess the water quality due to changing of land use (He, 2003).

Nhue is a small river with a length of 76 km meandering flows from North to the South through Hanoi city and Ha Nam province. The river starts from Liem Mac Bridge to take water from Red River at Hanoi and ends at Phu Ly city before joining with another river to make a bigger watershed. Area of Nhue watershed is about 107,530 ha covering 20030ha of Hanoi city , 67727 ha of Ha Tay province (now merged to Hanoi) and 19710 ha of Ha Nam province. The watershed size in wide is about 20km.

Nhue River is a branch of Red River and connected to each other through a bridge. When bridge is opened, fresh water from Red river fill Nhue river. Because of that the river can be re-clean. However, there are a big amount of pollutants from many point pollution sources spilt in recent years reduce self-clean ability and degrade water quality of the river.

Water quality became a hot issues and attracted many researches for a long time, for example, the research by Dang Duc Nhan et al. (2001) showed high level of DDT in the mud of river bed and in water; Nhue water quality project done by National center of natural sciences and technology with many valuable outputs (Trinh Anh Duc et al., 2006, Trinh Anh Duc et al., 2007 and Thi Thuy Duong, 2007); and water quality monitoring done by Environmental Protection Department. Most of the researches showed that Nhue water quality was heavily degraded with many parameters lower than the standard level for irrigation. Nevertheless, Nhue River is still the irrigation and drainage river for more than 100000 ha. Therefore it is valuable to have research to contribute on improving the water quality and environment.

Objectives of research are to study: the irrigation capacity of the river for crop production; degradation of river water quality; and accumulation of pollutants in the field.

2. Methodology

Maps of watershed

The river and its watershed was delineated on the GIS software with all layers of information of topography, administrative, soil, land use, irrigation and drainage systems.

Irrigation water

Irrigation water is calculated based on the water balance equation, mostly in the root zone layer (Pham Ngoc Hai et al., 2006)

$$\Sigma W = (ETc + W_s) - (W_o + \Sigma P_o + \Delta W) \tag{1}$$

$$W_s = A.\beta_k.H \text{ (mm)} \tag{2}$$

Ws is limited by condition of

$$W_{\beta min} \leq W_s \leq W_{\beta max} \text{ (mm)} \tag{3}$$

$$W_{max} = A.\beta_{max}.H; \text{ (mm)} \tag{4}$$

$$W_{min} = A.\beta_{min}.H \text{ (mm)} \tag{5}$$

$$W_o = A.\beta_o.H \text{ (mm)} \tag{6}$$

$$\Sigma P_o = \Sigma \alpha CP \text{ (mm)} \tag{7}$$

$$\Delta W = W_H + W_m \text{ (mm)} \tag{8}$$

$$W_H = A \, \beta_o.(H-H_{add}) \text{ (mm)} \tag{9}$$

$$C = 1-\sigma$$

where:

Σm: Total required water need for irrigation (mm)

ET_c: Evapotranspiration (mm)

W_s: Water need to be stored in the soil at the end of period (mm)

γ_k:Soil bulk density (ton m^{-3})

β_k : Soil moisture content at the end of period, analogue % of soil bulk density

W0: Available water at the begining of period

βmax, βmin: minimum and maximum soil moisture content

$W_{\beta max}$, $W_{\beta min}$: water at minimum and maximum soil moisture content (mm)

A: Soil porosity (%)

H: depth of root zone layer (mm)

ΣP_o: Water that plants use during the period:

P: Real precipitation (mm)

C: Fraction of precipitation go into the soil

σ: Flow factor in practice

α: Rainfall use factor

ΔW: The addition water that plant used during the period (mm)

W_H: Water that plant used thank to incrementof root (mm)

H_{add}: increment depth of root (mm)

W_m: Ground water that plant can use through capilary rise, this depends on ground water table and soil characteristics

Irrigation water for rice

Paddy rice is a case because rice growing in the flooding condition with a hardpan at the bottom of cultivated layer. Therefore there is a typical irrigation procedure for paddy rice. Irrigation water for rice was calculated based on the water requirement and water balance for whole crop duration and can be described as follow:

Water requirement for whole rice crop (M) in equation 1:

$$M = M1 + M2 \tag{10}$$

Where: M1: irrigation water for land preparation; M2: irrigation water for rice.
Irrigation water for land preparation is calculated in equation 11

$$M_1 = W_1 + W_2 + W_3 + W_4 - CR \tag{11}$$

$$W_1 = A.H(1-\beta_o) \tag{12}$$

$$W_3 = K \frac{H+a}{H}(t_a - t_b) \tag{13}$$

$$W_4 = ET_o.t \tag{14}$$

where:

W_1: water volume to saturate cultivated layer

W_2: water volume need to make a layer of water

W_3: Stable infiltration water in land preparation period
W_4: Evaporation during land preparation period
C: Rainfall effective factor
R: Rainfall (mm)
A: soil porosity
H: depth of soil cultivated layer (mm)
β_o: soil moisture content before irrigation (%)
H: depth of soil cultivated layer (mm)
t_a: Land preparation time (day)
t_b: Time to saturate the cultivated layer (day)
Water balance in the field is described in equation 15

$$W = (W + R + F_{in} + Cap) - (ETc + F_{out} + Per) - (S_{lim} - S_{ini}) - (G_{lim} - G_{ini}) \quad (15)$$

where
W: Current soil water
R: Rainfall in observed period
F_{in}: Inflow from higher fields
Cap: Capillary rise
ETc: Evapo-transpiration (ETc) calculated by Penman – Monteith method (FAO, 1998)
F_{out}: Surface outflow
Per: Percolation to ground water
S_{lim}; G_{lim}: limited surface and ground water due to using condition
S_{ini}; G_{ini}: Initial available surface and ground water
Irrigation will be set on when $\Delta < 0$ and vice versa

Water quality assessment

Water samples were taken at three locations along the river length, the upstream, middle and the downstream, in dry and rain seasons using a litter plastic sampler. A liter of sub-sample of sub-sample was collected from a mixed three sanmples at three depths of surface, middle and bottom. Samples were treated by HNO_3 0.1M and stored in a cool condition of 4 degree census. Water samples were analyzed pH_{H2O}, EC, COD, BOD_5, total phosphorus, total nitrogen, total E. coliform, Hg, Cd, Pb and Cu. Data were analyzed by SPSS software and compared with national water quality standard.

Estimation of pollutant mass

Pollution mass was estimated as a product of irrigation volume and concentration of pollutant at location, and time of irrigation

3. Results

3.1 Study area

Nhue River located at coodinate from $20^030'40''$ to $21^009'N$ and from $105^037'30''$ to $106^002'E$, within Red River Delta and bounded by Red River in the north, Day river in the west and Chau Giang river in the south. The river orginirated from Red River (Figure 1).

Topography

Whole watershed is located within Red River Delta with an elevation ranges from 1 to 10 m above sea level and lowered from north to south and from Red river, Day river to Nhue river.

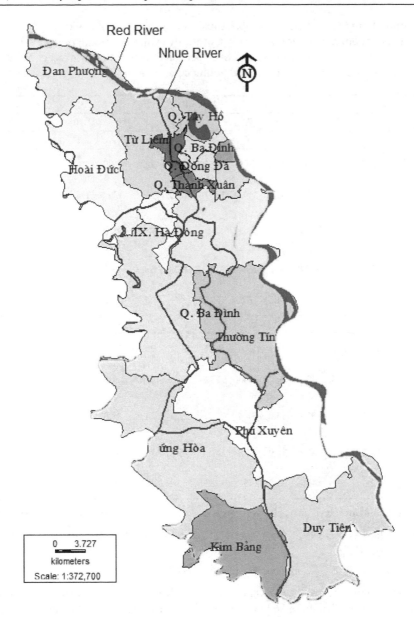

Fig. 1. Map of Nhue river and Nhue watershed with districts

Climate and hydrograph

The watershed is located in the sub-tropical region with 4 distinction seasons of spring, summer, autumn and winter. Average temperature is about 23 degrees. Annual rainfall in the watershed is from 1600 to 1700 mm (Table 1) and distributed unequally makes two different seasons, dry season from November to April and rain season from May to October. Very high rainfall concentrated from May to October causes exceeding water while very low rainfall in dry season, not enough water for some crops. Monthly evaporation ranges from 40 to 270 mm. In dry season, water balance can be negative because evaporation is exceeds rainfall while irrigation requirement for winter crops and young states of spring rice were very high.

	Month												Yearly
	1	2	3	4	5	6	7	8	9	10	11	12	
Phu Ly station													
Temperature (°C)	16.7	18.0	20.4	24.5	26.9	29.0	29.1	28.1	27.0	24.8	21.4	17.8	23.6
Rain fall (mm)	20.6	27.6	47.0	44.4	243.0	175.6	307.2	315.2	376.0	174.5	36.8	48.0	1815.9
Relative humidity (%)	85.4	90.4	89.4	90.2	87.4	83.0	84.0	74.0	88.0	84.3	83.5	83.8	85.3
Evaporation (mm)	63.4	42.3	59.8	54.7	69.4	271.2	100.1	59.8	63.9	89.1	90.6	76.1	1040.4
Sunshine hour (hour)	64.0	150.2	47.6	94.8	175.6	172.2	180.6	158.4	155.4	140.8	140.8	96.8	1577.1
Ha Dong station													
Temperature (°C)	16.8	16.6	20.7	24.3	27.0	29.0	29.0	28.1	27.1	25.0	21.5	17.9	23.6
Rain fall (mm)	18.0	28.2	43.6	59.2	226.4	250.8	317.8	328.2	196.6	87.8	30.0	29.0	1615.6
Relative humidity (%)	83.2	87.4	85.8	89.0	87.4	83.8	85.0	74.6	88.0	82.8	81.3	81.8	84.2
Evaporation (mm)	66.2	56.6	69.4	57.9	74.8	261.4	96.6	70.0	65.3	88.6	87.1	75.4	1069.3
Sunshine hour (hour)	62.0	46.6	45.4	83.8	145.4	142.4	157.6	145.2	148.6	136.3	137.0	97.3	1347.5

Table 1. Climate characteristic at Ha Dong (upper end) and Phu Ly station (lower end of river).

Ha Dong station is located at the upper end of the river Phu Ly station is located at the lower end of the river

River networks

Nhue River is bounded by three rivers: Red River, Day River and Chau River, flows from north to south with a distant of 76 km and has both functions of irrigation and drainage. The river takes water from Red river to irrigate for about 75-80% total water requirement of the basin and drainage for about 50-54% of the system

Ground water

Ground water in the basin strongly depends on the surface water sources, especially the water level in the Red River, the biggest river in the north and goes along the north east side of Nhue River. Ground water level and potential are fluctuated up and down as results from up and down of Red River water level. Ground water quality is also polluted because of receiving pollutants from surface water (ICEM, 2007).

Soil

Soil in the watershed can be classified as three main types of rusty spotted fluvisols distributed at the high locations, e.g. along the western parts with significant high content of ion and mangan that lateritic process happening, rusty spots can be formed with lower pH when dry; Eroded and impoverished soils distributed at the high level along Red River that can easily be eroded fine materials and nutrients; Red River old gleyed fluvisols distributed in the lower part in the south of the watershed that gley process can easily happen if drainage system is poor (Figure 2). The main characteristics of the soils are presented in Table 2

Soil properties	Soil types		
	Rusty spotted fluvisols	Eroded and impoverished	Old alluvial gleyed fluvisols
Soil depth (cm)	50	50	50
pH	5.5	5.5	6
OC (%)	2.04	1.72	2.43
Soil texture	Loam	Silt loam	Silty clay
Porosity (%)	46.3	50.1	46.4
Bulk density (kg cm^{-3})	1.32	1.34	1.30
Initial soil moisture content (%)	58	55	60
Infiltration Index	0.35	0.3	0.4
Initial infiltration rate (mm day^{-1})	145	72	32
Stable infiltration rate (mm day^{-1})	7.9	4.7	2

Table 2. Soil characteristics

Population

The population of the watershed is 4032884 in 2010 including 2527695 in the urban and 1505189 in the rural area. Very high population is in Hanoi capital with 2445861 (2194433 in the urban area and 251428 in the rural area). Because of quickly urbanization and industrialization in these provinces population of the watershed is increasing very fast. This increasing population strongly influences into the watershed water balance because a lot of surface water, ground water extraction and a huge amount of waste water contributed to the system, especially the big capital city with a lot of residential areas, industry, hospital, craft villages and agriculture activities (ICEM, 2007).

Land use

Detail land use types for three provinces are presented in Table 3. Dominated crop in the watershed is paddy rice (Figure 2) which grew in spring and summer season. Crops are grown in the rotation that can be a double crop as spring rice – summer rice; peanut – summer rice, can be a triple crop as spring rice – summer rice – winter maize; spring rice – summer rice – vegetable, can be year round non-rice annual crops (Table 4) as multi-vegetables that the farmer can have from three to seven harvests per year.

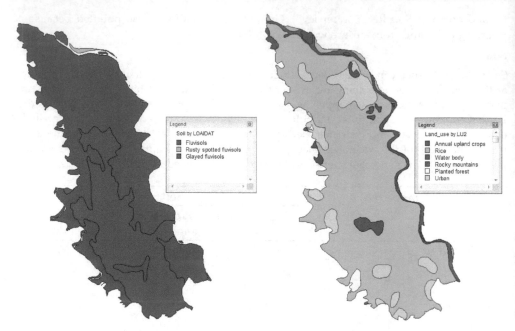

Fig. 2. Soil map (left) and land use map (right) of study area that many short non-rice annual crops hidden in the rice land use type

TT	Land use	Provincial and total area (ha)			
		Ha Noi	Ha Tay	Ha Nam	Total
	Natural area	25777.3	81315.1	22937.9	130030.35
1	Agricultural land	7958.1	44779.6	13091.4	65829.1
1.1	Agricultural production	5317.5	40140.7	10943.1	56401.4
1.1.1	Annual crops	4546.9	37361.7	9908.7	51817.4
1.1.1.1	Rice	2426.6	33188.1	9827.3	45441.9
1.1.1.2	Non-rice annual crops	2120.4	4173.6	281.5	6575.5
1.1.2	Perennial tree	770.6	2779.0	1013.6	4563.2
1.2	Forestry	0.0	0.0	0.0	0.0
1.3	Aquaculture	2508.5	3444.4	1808.1	7761.1
1.4	Salt production	0.0	0.0	0.0	0.0
1.5	Other agricultural lands	132.0	1194.5	1.0	1327.6
2	Residence and special land	17557.5	35335.5	9803.7	62696.8
3	Non-cultivated land	261.7	1200.0	42.8	1504.4

Sources: planning data from Hanoi city and Ha Nam province

Table 3. Main land uses in the watershed in 2010 (ha)

Sub-watershed	Irrigation unit	Province	District	Meteorology station	Spring rice (ha)	Maize (ha)	Soybean (ha)	Peanut (ha)	Sweet potato (ha)
1	Thanh Tri	Hanoi	Thanh Tri	Hanoi	1628	411	53	85	17
2	Hanoi center	Hanoi	NTHN	Hanoi	788	1365	0	0	0
3	Hong Van	Ha Tay	Thuong Tin	Hà Dong	6093	806	2255	161	262
4	La Khe	Ha Tay	TX Ha Dong	Hà dong	1841	88	44	41	74
			Thanh Oai	Ha Dong	6941	355	937	116	614
5	Phu Xuyen	Ha Tay	Phu Xuyen	Hung Yen	8544	742	8392	291	136
6	Ung Hoa	Ha Tay	Ung Hoa	Hung Yen	10869	827	3375	238	297
7	Dan Hoai Tu	Ha Tay	Dan Phuong	Hanoi	2049	1294	1602	2	155
		Ha Tay	Hoai Duc	Hanoi	3337	984	115	95	414
		Ha Tay	Tu Liem	Hanoi	1234	16	0	0	0
8	Huu Day	Ha Nam	Phu Ly	Phu Ly	1156	374	196	179	110
		Ha Nam	Kim Bang	Phu ly	5942	1730	514	337	338
		Hà Nam	Duy Tien	Phu Ly	6542	829	1014	277	181

Table 4. Distribution of annual crops in some districts in Nhue watershed

Current irrigation

Base on current topography and infrastructure, irrigation in the watershed can be classified as two types: irrigation by pumping to service of 16653 ha and semi pumping, that water can be free flow from the river to the field when river water level is higher than the field elevation and is pump to irrigate when river water level lower than the field elevation, with 55527 ha.

3.2 Irrigation
3.2.1 Irrigation for spring rice
Water requirements for spring rice are presented in Figure 3. Spring rice starts from the beginning of February and ends in the first and second weeks of June. The season before the spring rice season is the winter season that soil can be in many different land uses as fallow (for the double rice rotation, farmer plows the soil after harvesting summer rice); maize (in the rotations of rice-rice-maize), soybean, beans, and vegetables. The fallow soil is very dry before spring rice season but for triple rotation winter crops are harvested just before spring rice and soil still moist due to the irrigation during the crop. Because of that, soil moisture content is varied from soil to soil and crop to crop and a lot of irrigation water is needed for saturated and establish a water layer during land preparation before transplanting of spring rice.

Water balance and irrigation water were calculated and showed in Figure 4. The long dry period in the first half of season and high stable field water level require high irrigation in this period. Evapo-transpiration increases in the second half of season. ETc is highest at the flowering time, the last week of April and the first week of May when plant has maximum leaf area index. ETc is lowered in the maturing period until harvest time because of leaf elder, dying of the old leaf. However, rainfall in this period increases with some high rainfall events. As a result, irrigation is lowered with longer intervals. Figure 4 also shows that at some events with either high rainfall or high field water level. Water is succeeded and runs over the bunds (e.g. on the events of April 17, June 1 and June 14).

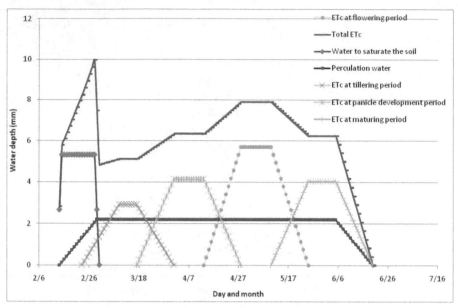

Fig. 3. Water losses for land preparation, percolation, and evapotranspiration of sping rice

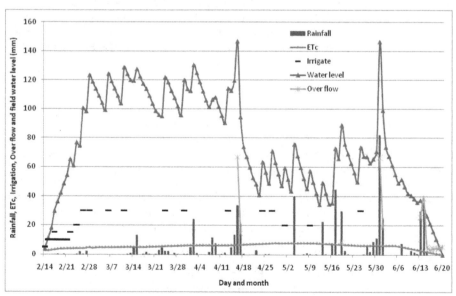

Fig. 4. Rainfall, ETc, irrigation water, field water level and overland flow for spring rice

Because summer rice is transplanted at an early in July, the intercrop time from harvesting spring rice to summer rice is shortness than a month. The longer this period the better for land preparation and decomposition of fresh organic matter from spring rice residue. Therefore, agronomists want to lengthen this period by shortening spring rice ripening period. Also, spring rice late ripening period, the time after silk ripening or about 10 days before harvesting, requires less and less water because of ending development. Farmer usually draws water out of the field in order to: shorten crop duration for harvesting on time; strengthen rice plant to avoid falling down in the heavy rain or storm/typhoon; harden the field surface easier for transportation (Pham Ngoc Hai et al., 2006).

3.2.2 Irrigation for summer rice

After harvesting spring rice in June land is plowed and harrowed immediately for summer rice. Because soil is already wetted and excessive rain water in this period, water for saturating the soil and creating flooded water layer is very low. Summer rice is transplanted in July and harvested at the end of September and early of October. This growing period is coin side with highest rainfall period of the year (about 30-40% of the annual rainfall during July-September period), soil is usually saturated and the field is usually flooded, because of that irrigation volume and intensity in summer rice is lower than the one in spring (Figure 5) rice.

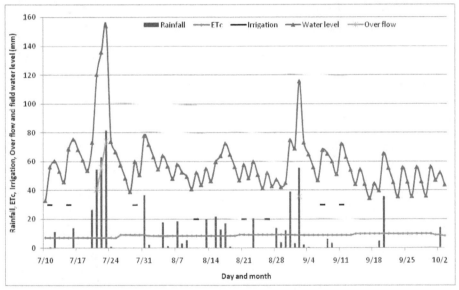

Fig. 5. Rainfall, ETc, irrigation water, field water level and overland flow for summer rice

3.2.3 Irrigation for non-rice annual crops

Non-rice annual crops or the upland crops, not planted in flood condition. However these crops need moist soil to grow.

Representative for these crops are soybean, sweet potato and maize grown in different seasons with different rainfall, Kc values and ETc (Figure 6). The condition for applying irrigation commands are soil moisture content drops lower than 65% of field capacity and soil moisture content should be best within 70 and 90% of field capacity. This parameter is measured by loggers at the hydrological stations, but in practice, it can be estimated and irrigation is considered after group discussion of water resource people, agronomist/extensionist and local management people.

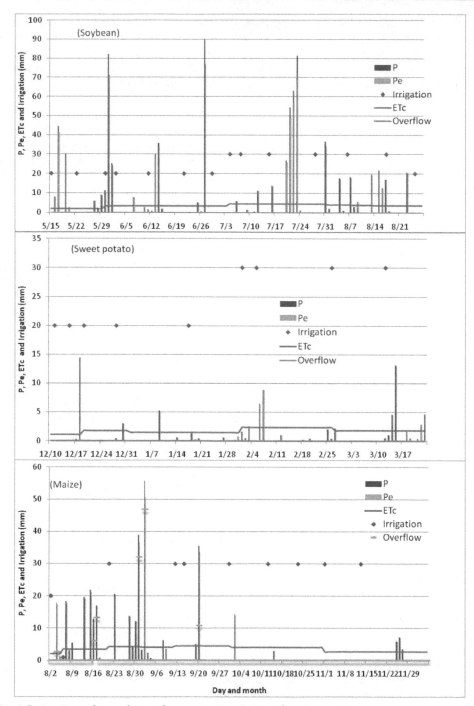

Fig. 6. Irrigation scheme for soybean, sweet potato and maize

3.2.4 Seasonal crop irrigation water

This water is summarized from all irrigation time during crop growing season, for example, total irrigation water of spring rice is summarized from water needed to saturate the soil, water needed to make a field water layer, for percolation, and water for evapo-transpiration. Summarizing the volume of irrigation water for whole crop is presented in Table 5.

Crop	Growing time	Water volume (m^3 ha^{-1})
Spring rice	Feb-Jun	4800
Summer rice	Jul-Oct	3400
Maize	Aug-Dec	2600
Sweet potato	Oct-Mach	2200
Soybean	May-Aug	1300

Table 5. Water needed for whole crop production (m^3)

3.3 Water quality

Physically, Nhue River takes fresh water from Red River, therefore, the river can re-clean itself after each irrigation time. However, the river flows through the capital and cities and many towns, villages, industrial factories, hospitals and craft villages containing strong point pollution sources. Especially, in recent two decades the increase of urbanization and industrialization in this basin rapidly increase the amount of pollutants to the river. Water quality becomes worse and worse (Trinh Anh Duc et al., 2006; Trinh Anh Duc et al., 2007; Thi Thuy Duong et al., 2007; and ICEM, 2007). The average daily discharge of waste water into the river is estimated as 2,554,000m^3 from agriculture and animal husbandry, 636,000 from industry, 619,000 from resident, and 15,500 from hospital, and especially waste water from To Lich river comes from the Hanoi city center with daily discharge of about 300,000-350,000m^3, excluding the disposal from cities and river side slowing down the river flow (ICEM, 2007). There is a big different in water quality in spatial and temporal. In dry season concentration of all elements are very high because very low fresh water comes from Red river that can not help to clean polluted water in the river. pH from Table 6 also shows the tends it increases in rain season as a result from dilution. Many pollutants have concentration higher than Vietnamese standard line of water quality, for example, pH, Coliform, total nitrogen, total P in dry season. However, measurements results showed that concentration of all heavy metals are lower than the standard line in both dry and wet seasons.

3.4 Estimation the mass of pollutants to the field

A mass of pollutants was calculated as product of total discharge and pollutant concentration. Pollutant concentration is complicated to determine because it varies spatially and temporally. It can be diluted by fresh water from Red River and rain water but it can also be more concentrated because of contribution of pollutants from many pollution sources as urban, industries, hospitals, craft villages and other activities. Dynamically, pollutants can easily be transported from upstream to downstream and always high concentration at the downstream of the watershed. Water in the river can be re-cleaned after rainfall but it will quickly be high concentration because of ever flows from pollution sources in industrialized and urbanized centers. In practice, irrigation is take place when only the soil is dried or long time after raining; it means the dilution effect is very small or insignificant. Because of that pollutant concentration for mass calculation in this study is distinguished as concentration in dry season and the one in rain seasons as presented in Table 6. Then the mass of pollutants will be calculated as product of total irrigation water and pollutant concentration (either in dry or rain season depend on the crop calendar, dry season from October to April and the rain season from May to September) and shown in Table 7.

Position	pH	EC* μS cm⁻¹	BOD₅ mgl⁻¹	COD mgl⁻¹	Coliform MPN/ 100ml	N_tot mgl⁻¹	P_tot mg l⁻¹	Cu mg l⁻¹	Pb mg l⁻¹	Hg mg l⁻¹	Cd mg l⁻¹
March 2010											
Lien Mac	7.5	4.15	9.6	22.4	110000	20.16	12.104	0.0154	0.0013	<LOD	0.00021
To bridge	7.31	6.46	70.4	51.2	460000	12.32	13.699	0.0058	0.0012	<LOD	0.00013
Nhat Tuu	7.17	5.32	16	26.4	1500	22.4	7.965	0.0192	0.0044	0.00013	<LOD
July 2010											
Lien Mac	8.94	178	4.8	8.6	90	4.81	0.60	0.0058	0.0012	<LOD	0.00013
To Bridge	10.56	252	32	56	210	5.6	0.86	0.0154	0.0013	<LOD	0.00021
Nhat Tuu	10.39	324	44	64	4600	7.20	1,35	0.0192	0.0044	0.00013	<LOD
TCVN: A	6-9		30	50	3000	5	4	2	0.1		0.005
B	5.5-9		50	80	5000	10	6	2	0.5		0.01

Note: March in dry season, July in rain season. Lien Mac, To bridge and Nhat Tuu are upstream, middle and downstream points along Nhue river, TCVN 5945-2005 is Vietnamese water standards for drinking (A) and irrigation (B). * Electrical conductivity

Table 6. Spatial and seasonal water quality of Nhue river

Position, crop	BOD₅	COD	Coliform	N_tot	P_tot	Cu	Pb	Hg	Cd
Lien Mac (upstream)									
Spring rice	46.08	107.52	5280000	96.768	58.0992	0.07392	0.00624	NS	0.001008
Summer rice	16.32	29.24	306	16.35	2.04	0.01972	0.00408	NS	0.000442
maize	12.48	22.36	234	12.506	1.56	0.01508	0.00312	NS	0.000338
Sweet potato	21.12	49.28	2420000	44.352	26.6288	0.03388	0.00286	NS	0.000462
Soybean	6.24	11.18	117	6.253	0.78	0.00754	0.00156	NS	0.000169
To (middle)									
Spring rice	337.92	268.8	1008	26.88	4.128	0.07392	0.00624	NS	0.001008
Summer rice	108.8	190.4	714	19.04	2.924	0.05236	0.00442	NS	0.000714
maize	83.2	145.6	546	14.56	2.236	0.04004	0.00338	NS	0.000546
Sweet potato	154.88	123.2	462	12.32	1.892	0.03388	0.00286	NS	0.000462
Soybean	41.6	72.8	273	7.28	1.118	0.02002	0.00169	NS	0.000273
Nhat Tuu (down stream)									
Spring rice	76.8	307.2	22080	34.56	6.48	0.09216	0.02112	0.000624	NS
Summer rice	149.6	217.6	15640	24.48	4.59	0.06528	0.01496	0.000442	NS
maize	114.4	166.4	11960	18.72	3.51	0.04992	0.01144	0.000338	NS
Sweet potato	35.2	140.8	10120	15.84	2.97	0.04224	0.00968	0.000286	NS
Soybean	57.2	83.2	5980	9.36	1.755	0.02496	0.00572	0.000169	NS

NS: not significant

Table 7. Mass of pollutant from irrigated water to the field (kg ha⁻¹season⁻¹) at upstream, in the middle and downstream positions

There are quite high mass of BOD_5, COD, coliform, total nitrogen, total phosphorus from Nhue river applied to the field in the watershed throughout irrigation as results from polluted sources. Adding of BOD_5, COD and coliform are the negative to the soil environment but nitrogen and phosphorus may have some positive effects because they provide more nitrogen and phosphorus nutrients to the soil for plant to grow. At some locations and time very high N and P was irrigated e.g. spring rice and sweet potato in the upstream of the river or rice, maize and potato in the downstream of the river where pollutants accumulated most of the time at the downstream of the river. There are about 0.02 to 0.09 kg ha^{-1} of Cu irrigated to the field in each crop (with highest amount in spring rice field in Nhat Tuu); from 0.002 to 0.02 kg ha^{-1} of Pb irrigated to the field (with highest amount of 0.02 in spring rice in Nhat Tuu; amount of irrigated Hg is insignificant at the upstream and middle but significant at the downstream. In contrast to Hg, amount of Cd irrigated to the field is significant at the upstream and middle but insignificant at the downstream. Because heavy metal concentration is lower than standard line for irrigation water, so that its amount can be considered that safe to the environment. However, the long term accumulation of these elements can be harmful to the soil, environment, product and people.

5. Conclusions

Nhue river is one of the important river and watershed in Vietnam because it irrigates and drains for more than 100000 ha of agricultural land of three provinces and drains most of the waste water for Hanoi capital and two other provincial towns with hundred of industrial, urban area, hospital, craft villages and intensive agriculture. From research results we can draw some conclusion follow:

Water quality of the river is degrading very strong with high content of BOD_5, COD and Coliform that threatening soil and water quality and people health of the whole watershed. Heavy metal concentration in the river water is still lower than the standard line. However, river water strongly impact on environment because it relates to people activity and agricultural production. Mass pollutant calculation showed that on the other side of positive effects of providing nitrogen and phosphorus from polluted irrigation water there are many negative impacts of long term accumulation of BOD, COD, coliform and heavy metal in the soil and water that can be serious problems in the future. One of the reason is there are many continuous pollution sources from urban and industry to release pollutants to the river water and spreading to the field causing soil and water degradation. The Government should have strong action to control the pollution sources and clean the environment for sustainable development.

6. References

Dang Duc Nhan, F.P. Carvalho, Nguyen Manh Am, Nguyen Quoc Tuan, Nguyen Thi Hai Yen, J.-P. Villeneuve, C. Cattini, 2001, Chlorinated pesticides and PCBs in sediments and molluscs from freshwater canals in the Hanoi region, *Environmental Pollution*, Vol.112, pp. 311- 320

FAO, 1998, Crop evapotranspiration - Guidelines for computing crop water requirements - *FAO Irrigation and drainage paper 56*. Food and Agriculture Organization of the United Nations, Rome.

Gardi, C. (2001). "Land use, agronomic management and water quality in a small Northern Italian watershed." *Agriculture, Ecosystems & Environment, Vol.* 87 No.1, pp. 1-12

He, C. (2003). "Integration of geographic information systems and simulation model for watershed management." *Environmental Modelling & Software* Vol. 18, No. 8-9, pp. 809-813.

ICEM, 2007, Day/Nhue river basin pollution sources study. Improving Water Quality in the Day/Nhue River Basin, Vietnam: Capacity Building and Pollution Sources Inventory, ministry of natural resources and environment, 155 pp.

Karr, J. D., W. J. Showers, et al. (2003). "Low-level nitrate export from confined dairy farming detected in North Carolina streams using 15N." *Agriculture, Ecosystems & Environment*, Vol. 95 No.1, pp. 103-110.

Kraft, G. J. and W. Stites (2003). "Nitrate impacts on groundwater from irrigated-vegetable systems in a humid north-central US sand plain." *Agriculture, Ecosystems & Environment*, Vol.100, No.1, pp. 63-74.

McDowell, R. W., A. N. Sharpley, et al. (2003). "Modification of phosphorus export from an eastern USA catchment by fluvial sediment and phosphorus inputs." *Agriculture, Ecosystems & Environment* Vol. 99, No. 1-3, pp. 187-199.

Nikolaidis, N. P., H. Heng, et al. (1998). "Non-linear response of a mixed land use watershed to nitrogen loading." *Agriculture, Ecosystems & Environment, Vol.* 67, No.2-3, pp. 251-265.

Pham Ngoc Hai, Tong Duc Khang, Bui Hieu and Pham Viet Hoa, 2006, irrigation and irrigation regimes for crops, *plan and projection of irrigation systems*, Construction Publishing House, pp. 46-93 (in Vietnamese)

Thi Thuy Duong, Agne`s Feurtet-Mazel, Michel Coste, Dinh Kim Dang, Alain Boudou, 2007, Dynamics of diatom colonization process in some rivers influenced by urban pollution (Hanoi, Vietnam), *Ecological Indicators* Vol. 7, pp. 839–851

Trinh Anh Duc, Marie Paule Bonnet, Georges Vachaud, Chau Van Minh, Nicolas Prieur, Loi Vu Duc, Le Lan Anh, 2006, Biochemical modeling of the Nhue River (Hanoi, Vietnam): Practical identifiability analysis and parameters estimation, *Ecological Modelling*, Vol. 193, pp. 182-204.

Trinh Anh Duc, Georges Vachaud, Marie Paule Bonnet, Nicolas Prieur, Vu Duc Loi, Le Lan Anh, 2007, Experimental investigation and modelling approach of the impact of urban wastewater on a tropical river; a case study of the Nhue River, Hanoi, Viet Nam, *Journal of Hydrology*, Vol. 334, pp. 347– 358

Zhang, J. and S. Erik Jorgensen (2005). "Modelling of point and non-point nutrient loadings from a watershed." *Environmental Modelling & Software* Vol. 20, No. 5, pp. 561-574.

5

Recycling Vertical-Flow Biofilter: A Treatment System for Agricultural Subsurface Tile Water

K.H. Baker[1] and S.E. Clark[2]
[1]*Life Science Program, Penn State Harrisburg*
[2]*Environmental Engineering Programs, Penn State Harrisburg*
USA

1. Introduction

Agricultural runoff and similar nonpoint sources of pollution are responsible for widespread degradation of surface water quality in the U.S. (Hall and Killen, 2005; Hardy and Koontz, 2008). In almost three-quarters of the rivers studied in the National Water Quality Survey, nonpoint discharges were major contributors to water quality impairment (US EPA, 1992). Nonpoint source discharges resulting from agricultural runoff add large amounts of inorganic nitrogen and phosphorus to surface water (Goolsby and Battaglin, 2001; Powers, 2007; US EPA, 1992). In the Chesapeake Bay Region (US), nonpoint source discharges contribute about two-thirds of the nitrogen and one-quarter of the phosphorus inputs (Correll *et al.*, 1995). In the 1200 km[2] Conestoga River watershed in Pennsylvania, 47.2 kg/ha/yr total nitrogen and 44.7 kg/ha/yr nitrate-nitrogen are discharged from nonpoint sources adding, ultimately, to the nutrient load of the Chesapeake Bay (Woltenmade, 2005). The addition of excessive inorganic nutrients to surface waters leads to eutrophication, which, in turn, is associated with the development of hypoxic zones such as those in the Gulf of Mexico, the Chesapeake Bay, and similar areas (Alexander *et al.*, 2008; Boesch *et al.*, 2001; Mitsch *et al.*, 1999; Wang *et al.*, 2001).

Subsurface tile drainage is a common agricultural water management practice used in regions with a seasonally high water table. By taking advantage of this system, farmers are able to extend their growing season by allowing for earlier spring planting and later harvest dates. The use of subsurface tile drainage has been shown to significantly improve crop production (Kladivko *et al.*, 2005). Skaggs *et al.* (1994) noted that subsurface artificial drainage has improved agricultural production on nearly one-fifth of U.S. soils. In the intensively cropped watersheds of the Midwest United States, the use of subsurface tile drainage has allowed one of the highest agricultural productivities in the world. Approximately 30% of all agricultural lands in the upper Midwest are artificially drained (Zucker and Brown, 1998).

Despite all of the benefits to crop production, tile drain lines can have a negative environmental impact. Tile drain lines can act as conduits for contaminants, promoting the rapid movement of these substances to surface waters (Fleming and Ford, 2004; Gentry *et al.*,

2000). A study of tile drain outlets in southwestern Ohio found an average concentration of nitrate-N of 17 mg L^{-1} was discharged to receiving waters (Fleming et al., 1998). The crop production system employed, the amount, rate, and timing of fertilizer application, the size and arrangement of drainage tiles, and the presence of cover crops are all known to influence nitrogen inputs to surface waters from tile drainage systems (Kaspar et al., 2007; Kladivko et al., 2004; Nangia et al., 2008; deVos et al., 2000; Domagalski et al., 2008; Dinnes et al., 2002).

Agronomic controls such as crop and fertilizer management, however, are not usually sufficient to rectify nutrient pollution resulting from tile drainage systems (Jaynes et al., 2008; Madramootro et al., 2007). Therefore, additional methods for nutrient removal and control are needed where subsurface tile drainage is common.

Passive treatment systems such as vegetated riparian zones and biofilters have been shown to be effective in controlling nutrient inputs from surface runoff (Cors and Tychon, 2007; Dodds and Oakes, 2006; Mankin et al., 2007; Mayer et al., 2007; Spruill, 2004; Yamada et al., 2007). The preponderance of water flow in tile drainage systems though, is subsurface, within the vadose zone. Therefore, the efficiency of surface systems for treatment may be reduced because substantial amounts of contaminated water may bypass the active treatment zone. To address this limitation, subsurface systems such as in-situ bioreactors, permeable reactive barriers, biofilters, and subsurface flow constructed wetlands have been investigated (Bezbaruh and Zhang, 2003; Darbi et al., 2003; Greenan et al., 2006; Robertson et al., 2007; Schipper and Vojvodic-Vakovic, 2000; Schipper and Vojvodic-Vakovic, 2001; Schipper et al., 2004; Su and Puls, 2007; van Driel et al., 2006). These subsurface systems generally depend on microbial denitrification to mineralize and remove nitrate. Denitrification is an anaerobic process. As such, it requires anaerobic conditions in the subsurface as well as an adequate supply of electron donors and available carbon. Thus, an exogenous source of carbon such as wood chips or sawdust is usually required for these systems to function properly (Greenan et al., 2006; Lin et al., 2002; Vymazal, 2007).

We recently reported on a recycling vertical-flow bioreactor (RVFB) for the treatment of household greywater (Gross et al., 2007). This system intercepts and re-circulates contaminated water to a vegetated soil biofilter for aerobic treatment. We report here on the potential use of this system for the removal of excess nutrients from tile water.

2. Materials and methods

The mesocosm scale RVFBs used in this study have been described elsewhere (Gross et al., 2007). Briefly, each unit consisted of two tiers, each made of a 55 x 40 x 30 cm plastic container. The top container functioned as a soil based treatment unit, while the bottom container served as a reservoir from which water was recycled continuously to the treatment unit (Figure 1).

The RVFB units were run with a total of 40 L of synthetic tile water (STW: [g L^{-1}] $CaCl_2$, 1.7; $NaSO_4$, 1.8; $NaHCO_3$, 0.1; KNO_3, 0.1; K_2HPO_4, 0.0004; Humic acid, 0.003) in a semi-batch mode. After initial loading of each unit, water from the reservoir was recirculated to the treatment unit at a rate of 0.41 L min^{-1} using a 4.6 L min^{-1} (5 watt) submersible pump. Milk tubing (0.635 cm) modified into a drip line provided uniform water distribution over the surface of the treatment unit. The recirculation rate was set to prevent ponding of water on the soil surface.

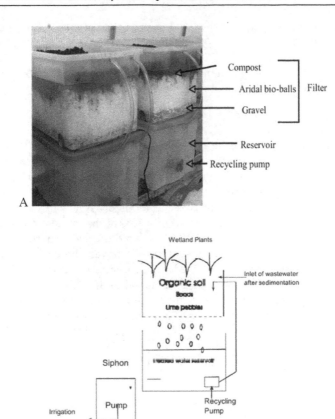

RVFB units used in this research. A. RVFB units at initial set-up of experiment. Note, no vegetation had been planted at this time. B. Schematic of RVFB.

Fig. 1. Recycling Vertical Flow Bioreactors (RVFB)

Evaluation of the RVFB was conducted under conditions typical of a temperate climate such as that found in the Mid-Atlantic region of the Eastern US. Experimental systems contained a mixed plant community of emergent plants common to southeastern Pennsylvania. An initial period of 4 weeks (designated weeks -4 to 0; data not shown) was allocated for the establishment of the plant community before STW was added for treatment. Plants were excluded from the control system by weeding twice per week. In addition, plants were harvested from one of the experimental treatments at day 30 of the growing season to assess the importance of vegetation in nutrient removal. The systems were maintained in a greenhouse at ambient temperatures for the duration of the study, one growing season (May – October).

After allowing for the initial plant establishment and system acclimation, samples were collected twice weekly by draining 20 L of effluent from the reservoir container of each RVFB and replacing it with 20 L of freshly prepared STW. An aliquot (1 L) of the drained effluent was transported on ice immediately to the laboratory for analysis. Samples were stored at 4°C and analyzed within 24 hours of collection. Nitrate-nitrogen (NO_3-N), nitrite-nitrogen (NO$_2$-N), and dissolved reactive (ortho) phosphate were assayed using standard chemical test kits (HACH Test-N-Tube Plus Methods 835/836, 839 and 834 respectively. All kits follow USEPA approved methods SM 4500. Hach Company, Loveland, CO.).

Data was analyzed using the statistical program Prism 4.0 (GraphPad, Inc). Treatments were compared using paired T-tests at a level of significance (α = 0.05)

3. Results

Figure 2 summarizes the removal of nitrate in the RVFB systems. There was no noticeable removal of NO$_3$-N from the influent tile water in the control system, indicating that passive removal via adsorption to the soil or microbial transformation was not a significant factor. Effluent nitrate concentrations in the vegetated systems were consistently below the EPA guidelines of 10 mg L^{-1}. In fact, effluent concentrations of NO$_3$-N in these systems rarely exceeded 2 mg L^{-1}, corresponding to a removal of > 90% of the influent NO$_3$-N. Harvesting of the plant community (day 30) resulted in a rapid increase in the effluent NO$_3$-N concentration. Within two weeks of the vegetation removal, the concentration of nitrate discharged by the harvested unit approached the concentration discharged in the control system indicating that the bulk of the nitrogen removal in the RVFB was the result of plant uptake and assimilation rather than of denitrification or other soil microbial processes. Nitrate levels in the effluent from the harvested unit remained significantly elevated, in excess of discharge limits, for the remainder of the study. The rapid increase in NO$_3$-N seen upon the removal of vegetation indicates that possible harvesting of the plants in a functioning RFVB should be limited to times when tile water discharge is minimal.

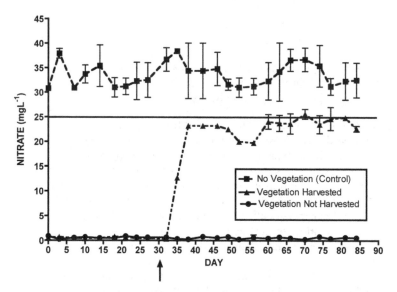

Mean +/- S.D. of effluent nitrate concentration in RVFB treating synthetic tile water (25 mgL^{-1} NO$_3$-N initial concentration: solid line). Plants were harvested in the 30th day of the experiment (arrow).

Fig. 2. Nitrate (NO$_3$-N) Removal in an RVFB

Ammonia-N in the effluent of all of the planted systems was consistently below 0.1 mg L^{-1} (data not shown). The concentration of NO_2-N was significantly ($p = 0.05$) lower in the vegetated units than in the control unit. (data not shown). Removal of vegetation from one of the vegetated units (Harvested; day 30) did not have a clear impact on the discharge of nitrite by that unit. While there appeared to be a slight elevation in the concentration of NO_2-N in the Harvested unit, this increase was transient and may not have been significantly different from the discharge of the remaining experimental units.

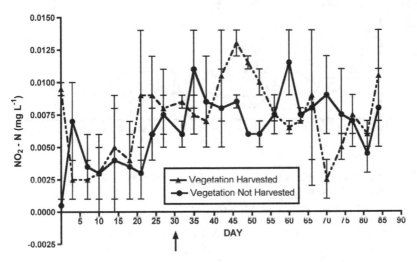

Mean +/- S.D. of effluent nitrite-N concentration in RVFB treating synthetic tile water. Plants were harvested in the 30th day of the experiment (arrow).

Fig. 3. Nitrite (NO_2-N) in RVFB Effluent

Removal of reactive phosphate from influent tile water was negligible in the RVFB without vegetation (Control; Figure 4). In fact, at times, the concentration of phosphate in the effluent was higher than that in the influent indicating that phosphate was being leached from the soil. In the presence of vegetation, the concentration of phosphate in the effluent was reduced to 20% of the control effluent. Removal of vegetation had no apparent impact on the concentration of PO_4 in the effluent. Retention of phosphate in the remaining root biomass as well as uptake by plant re-growth may account for the low concentrations of PO_4 in the post-harvest effluent, however, the specific cause of this pattern was not established.

4. Conclusions and future recommendations

Nutrient enrichment from non-point source runoff is a major factor in the degradation of surface water quality (Hardy and Koontz 2008, Ribaudo *et al.* 2001). The sources of nutrient runoff vary in scale from small individual households to large regional agricultural activities. Similarly, the options for prevention and remediation available for this type of pollution vary widely. Ultimately, multiple technologies at multiple scales must be available to address this issue (Ribaudo *et al.* 2001).

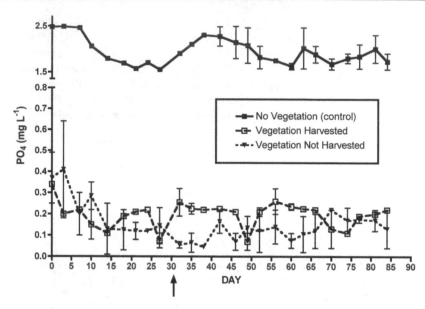

Mean +/- S.D. of effluent phosphate (PO₄) concentration in RVFB treating synthetic tile water. Plants were harvested in the 30th day of the experiment (arrow).

Fig. 4. Reactive (ortho) Phosphate in RVFB Effluent

The efficacy of management and control procedures in minimizing the runoff of nutrients has been documented widely (Mitsch *et al.*, 1999). However, agronomic practices alone are not enough to eliminate all agriculture-related nutrient runoff. In order to adequately reduce the impact of non-point source runoff, a combination of agronomic practices and treatment techniques is required.

A wide variety of treatment options for nutrient runoff have been developed. These include riparian buffer zones, biofilters, denitrification walls and constructed wetlands (Bezbaruah and Zhang, 2003; Darbi *et al.*, 2003; Jaynes *et al.*, 2008; Kelly *et al.*, 2007; Lin *et al.*, 2007; Su and Puls, 2007; Vymazal, 2007; Yamada *et al.*, 2007). Although each of these systems has been shown to reduce the NO_3-N concentration in groundwater, there is no single ideal system appropriate for use under all circumstances.

Our research demonstrates the potential use of a recycling vertical-flow biofilter (RVFB) as an alternative treatment option for the removal of nutrients from contaminated tile water. Using this relatively simple system we were able to achieve a > 95% removal of reactive phosphate and a > 90% removal of nitrate-nitrogen from STW.

The RVFB combines characteristics of a constructed wetland, a riparian buffer zone and a trickling filter for aeration. Subsurface flow intercepted by the RVFB is recycled to the soil surface. From there, it flows through a vegetated soil bed treatment system where combined biotic and abiotic processes remove excess nutrients and then flows through a layer of hollow plastic spheres, finally trickling into a reservoir. Movement of the water through the hollow spheres re-aerates the water and prevents the development of anaerobic conditions. Thus the RVFB is a hybrid treatment system combining the advantages of several existing treatment processes into one system capable of treating both surface and subsurface runoff.

For example, reactive phosphate is most likely removed by a combination of plant uptake and soil sorption. Removal of nitrogen compounds, on the other hand, is most likely the result of plant uptake as well as limited nitrification and denitrification by the soil microbial community.

The importance of plant uptake in the removal of nutrients is reflected in the increase in the concentration of both nitrate and phosphate in the harvested unit after the removal of vegetation from the system. No such increase in the concentration of nutrients was seen in units from which the vegetation was not removed. Because of the dominant role of vegetation in the RVFB, application of this technology to field situations must consider the management and ultimate use of the vegetation.

The RVFB has several advantages over other systems for the treatment of agricultural runoff. Since it is based on modular components – the upper soil treatment module, functionally similar to a riparian buffer, the plastic spheres, functionally similar to a trickling filter and the lower recirculation and reservoir module – there is flexibility in the design, allowing complete units to be tailored to a specific site. In addition, the use of separate modules should reduce maintenance costs since repair to a single component can be done by simply replacing the module without the need to disassemble the entire system and disrupting its operation.

In exploiting the ability of vegetated soil systems to sequester and transform inorganic nutrients, the RVFB reflects the advantages of a riparian buffer, a well-established treatment modality for the prevention of surface water pollution from agricultural runoff (Cors and Tychon, 2007; Mankin *et al.*, 2007; Mayer *et al.*, 2007; Schoonover *et al.*, 2005; Yamada *et al.*, 2007). By adding a subsurface recirculating reservoir, the RVFB also is capable of intercepting and treating subsurface runoff, particularly tile water that normally bypasses riparian buffers. Unlike constructed riparian wetlands, however, the RVFB operates in a primarily aerobic mode: water trickling through the soil and into the reservoir is aerated. Thus, the generation and release of nitrogenous greenhouse and ozone depleting gases and precursors of acidic deposition (e.g. N_2O) associated with denitrification-based systems may be avoided (Magner *et al.*, 2004; David *et al.*, 2009)

Our results provide a proof-of-concept only for the RVFB. Additional research is needed to demonstrate how well the system functions under realistic field conditions as well as the costs of this system compared to alternatives. We believe, however, that the RVFB has the potential to be a useful addition to the armamentarium in the fight against non-point source pollution.

5. Abbreviation list

RVFB: recycling vertical-flow bioreactor
STW: synthetic tile water

6. References

Alexander, R.B., R.A. Smith, G.E. Schwarz, E.W. Boyer, J.V. Nolan, and J.W. Brakebill. 2008. Differences in phosphorus and nitrogen delivery to the Gulf of Mexico from the Mississippi River Basin. Environ. Sci. Technol. 42:822 - 830.

Bezbaruah, A.N., and T.C. Zhang. 2003. Performance of a constructed wetland with a sulfur/limestone denitrification section for wastewater nitrate removal. Environ. Sci. Technol. 37:1690 - 1697.

Boesch, D.F., R.B. Brinsfield, and R.E. Magnien, 2001. Chesapeake Bay eutrophication: scientific understanding, ecosystem restoration, and challenges for agriculture. J. Environ. Qual. 30:303 - 20.

Correll D.L., T.E. Jordan, and D.E. Weller. 1995. The Chesapeake Bay watershed: effects of land use and geology on dissolved nitrogen concentrations. Pages 639-648 in P. Hill and S. Nelson, eds. Towards a sustainable coastal watershed: The Chesapeake experiment. Chesapeake Research Consortium, Solomons, MD.

Cors M. and B. Tychon. 2007. Grassed buffer strips as nitrate diffuse pollution remediation tools: management impact on the denitrification enzyme activity. Wat. Sci, Technol. 55:25 - 30.

Darbi A., T. Viraraghavan, R. Butler, and D. Corkal. 2003. Pilot-scale evaluation of select nitrate removal technologies. J. Env. Hlth. Sci. A. Tox. Hazard Subst. Environ. Eng 38:1703 - 1715.

David, M.B., S.J. DelGrosso, X. Hu, E.P. Marshall, G.F. McIsaac, W.J. Parton, C. Tonitto, and M.A. Youssef. 2009. Modeling denitrification in a tile-rained, corn and soybean agroecosystem of Illinois, USA. Biogeochem. 93:7-30.

deVos J.A., D. Hesterberg, P.A.C. Raats. 2000. Nitrate leaching in a tile-drained silt loam soil. Soil Sci. Soc. Am. J. 64:517 - 527.

Dinnes, D.L., D.L. Karlen, D.B. Jaynes, T.C. Kaspar, J.L. Hatfield, T.S. Colvin, and C.A. Cambardella. 2002. Nitrogen management strategies to reduce nitrate leaching in tile-drained Midwestern soils. Agronomy J. 94:153 - 171.

Dodds, W.K. and R.M. Oakes. 2006. Controls of nutrients across a prairie stream watershed: land use and riparian cover effects. Env. Manag. 37:634 - 646.

Domagalski, J.L., S. Ator, R. Coupe, K. McCarthy, D. Lampe, M. Sandstrom, and N. Baker. 2008. Comparative study of transport processes of nitrogen, phosphorus, and herbicides to streams in five agricultural basins. USA. J. Environ. Qual. 37:1158 - 1169.

Fleming, R. and R. Ford. 2004. Suitability of "end-of-pipe" systems to treat farm tile drainage water. Ridgetown College-University of Guelph. http://www.ridgetownc.on.ca/Research/research_fleming3.cfm. Last accessed 06/19/08.

Fleming, R., M. MacAlpine, and C. Tiffin. 1998. Nitrate levels in soil, tile drainage water and shallow groundwater under a variety of farm management systems. CSAE Paper 98 - 101, Vancouver, B.C.

Gentry, L.E., M.B. David, and K.M. Smith-Starks. 2000. Nitrogen fertilizer and herbicide transport from tile drained fields. J. Environ. Qual. 29:232 - 240.

Goolsby, D.A. and W.A. Battaglin. 2001. Long-term changes in concentration and flux of nitrogen in the Mississippi River Basin, USA. Hydrol. Proc. 15:1209 - 1226.

Greenan, C.M., T.B. Moorman, T.C. Kaspar, T.B. Parkin, and D.B. Jaynes. 2006. Comparing carbon substrates for denitrification of subsurface drainage water. J. Environ. Qual. 35:824 - 829.

Gross, A., D. Kaplan, and K. Baker. 2007. Removal of chemical and microbiological contaminants from domestic greywater using a recycled vertical flow bioreactor (RVFB). Ecol. Eng. 31:107 - 114.

Hall, L.W. Jr. and W.D. Killen. 2005. Temporal and spatial assessment of water quality, physical habitat, and benthic communities in an impaired agricultural stream in

California's San Joaquin Valley. J. Env. Sci. Hlth. A Tox. Hazard. Sub. Environ. Eng. 40:959 - 989.

Hardy, S.D. and T.M. Koontz. 2008. Reducing nonpoint source pollution through collaboration: policies and programs across the US states. Environ. Manag. 41:301 - 310.

Jaynes, D.B., T.C. Kaspar, T.B. Moorman, and T.B. Parkin. 2008. In situ bioreactors and deep-drain pipe installation to reduce nitrate losses in artificially drained fields. J. Environ. Qual. 37:429 - 436.

Kaspar, T.C., D.B. Jaynes, T.B. Parker, and T.B. Moorman. 2007. Rye crop cover and garnagrass strip effects on NO3 concentration. J. Environ. Qual. 36: 1503 – 1511.

Kelly, J.M., J.L. Kovar, R. Sokolowsky, and T. Moorman. 2007. Phosphorus uptake during four years by different vegetation cover types in a riparian buffer. Nutr. Cycl. Agroecosyst. 78:239 -251.

Kladivko, E.J., J.R. Frakenberger, D.B. Jaynes, D.W. Meek, B.J. Jenkinson, and N.R. Fausey. 2004. Nitrate leaching to subsurface drains as affected by drain spacing and changes in crop production system. J. Environ. Qual. 33(5):1803-1813.

Kladivko, E.J., G.L. Willoughby, and J.B. Santini. 2005. Corn growth and yield response to subsurface drain spacing on Clermont Silt Loan Soil. Agron. J. 97:1419 - 1428.

Lin, Y.F., S.R. Jing, T.W. Wang, and D.Y. Lee. 2002. Effects of macrophytes and external carbon sources on nitrate removal in constructed wetlands. Environ. Pollut. 119:413 - 420.

Lin, Y.F., S.R. Jing, D.Y. Lee, Y.F. Chang, and K.C. Shih. 2007. Nitrate removal and denitrification affected by soil characteristics in nitrate treatment wetlands. J. Environ. Sci. Hlth. A. Tox Hazard Subst. Environ. Eng. 42:471 - 479.

Madramootoo, C.A., W.R. Johnston, J.E. Ayars, R.O. Evans, and N.R. Fausey. 2007. Agricultural drainage management, quality and disposal issues in North America. Irrig. and Drain. 56:535 - 545.

Magner, J.A., G.A. Payne, and L.J. Steffen. 2004. Drainage effects on stream nitrate-N and hydrology in South-central Minnesota (USA). Environ. Monitor. Assess. 91:183-198.

Mankin K.R., D.M. Ngandu, C.J. Barden, S.L. Hutchinson, and W.A. Geyer. 2007. Grass-Shrub riparian buffer removal of sediments, phosphorus, and nitrogen from simulated runoff. J. Am. Wat. Res Assoc. 43:1108 - 1116.

Mayer, P.M., S.K. Reynolds jr., M.D. McCutchen, and T.J .Canfield. 2007. Meta-analysis of nitrogen removal in riparian buffers. J. Environ. Qual. 36:1172 - 1180.

Mitsch, W.J., J.W. Day Jr., J.W. Gilliam, P.M. Groffman, D.L. Hey, G. Randall, and N. Wang. 1999. Reducing Nutrient Loads, Especially Nitrate-Nitrogen, to Surface Water, Groundwater, and the Gulf of Mexico. Topic 5 Report for the Integrated Assessment on Hypoxia in the Gulf of Mexico. U.S. Dept. of Commerce, National Oceanic and Atmospheric Administration, Silver Spring, MD 111 p.

Nangia, V., P.H. Gowda, D.J. Mulla, and G.R. Sands. 2008. Water quality modeling of fertilizer management impacts on nitrate losses in tile drains at the field scale. J. Environ. Qual. 37:296 -307.

Powers, S.E. 2007. Nutrient loads to surface water from row crop production. Int. J. Life Cycle Assess. 12:399 – 407.

Ribaudo, M.O., R. Heimlich, R. Claassen, and M. Peters. 2001. Least-cost management of nonpoint source pollution: source reduction versus interception strategies for controlling nitrogen loss in the Mississippi Basin. Ecol. Econ. 37:183 - 197.

Robertson, W.D., L.J. Ptacek, and S.J. Brown. 2007. Geochemical and hydrological impacts of a wood particle barrier treating nitrate and perchlorate in ground water. Ground Wat. Monit. Remed. 27:85 - 95.

Schipper, L.A., and M. Vojvodic-Vukovic. 2000. Nitrate removal from groundwater and denitrification rates in a porous treatment wall amended with sawdust. Ecol. Eng. 14:269 - 278.

Schipper, L.A., and M. Vojvodic-Vukovic. 2001. Five years of nitrate removal, denitrification, and carbon dynamics in a denitrification wall. Wat. Res. 35:3473 - 3477.

Schipper, L.A., G.F. Barkle, J.C. Hadfield, M. Vojvolic-Vukovic, and C.P. Burgess. 2004. Hydraulic constraints on the performance of a groundwater denitrification wall for nitrate removal from shallow groundwater. J. Contam. Hydrol. 69:263 - 279.

Schoonover, J.E., K.W.J. Willard, J.J. Zaczek, J.C. Mangun and A.D. Carver. 2005. Nutrient attenuation in agricultural surface runoff by riparian buffer zones in Southern Illinois, USA. Agroforestry Systems 64:169-180.

Skaggs, R.W., M.A. Brevé, and J.W. Gilliam. 1994. Hydrologic and water quality impacts of agricultural drainage. Crit. Rev. Environ. Sci. Technol. 24:1 - 32.

Spruill, T.B. 2004. Effectiveness of riparian buffers in controlling ground-water discharge of nitrate to streams in selected hydrogeological settings of the North Carolina Coastal Plain. Wat. Sci. Technol. 49:63 - 70.

Su, C.M. and R.W. Puls. 2007. Removal of added nitrate in the single, binary, and ternary systems of cotton burr compost, zerovalent iron, and sediment: Implications for groundwater nitrate remediation using permeable reactive barriers. Chemosphere 67:1653 - 1662.

US EPA 1992. The national water quality inventory. The 1992 report to Congress. US EPA, Washington, DC.

Van Driel, P.W., W.D. Robertson, and L.C. Markley. 2006. Upflow reactors for riparian zone denitrification. J. Environ. Qual. 35:412 - 420.

Vymazal, J. 2007. Removal of nutrients in various types of constructed wetlands. Sci. Total Environ. 380:48 - 65.

Wang, P, R. Batiuk, L. Linker, and G. Shenk. 2001. Assessment of best management practices for improvement of dissolved oxygen in Chesapeake Bay estuary. Wat. Sci. Technol. 44:173 -80.

Woltemade, C. 2005. Potential for treatment wetlands to reduce non-point source nitrogen loads on a watershed scale: Modeling the Conestoga River watershed, Pennsylvania, USA. Geophys. Res. Abst. 7.00344.

Yamada, T., S.D. Logsdon, M.D. Tomer, and M.R. Burkart. 2007. Groundwater nitrate following installation of a vegetated riparian buffer. Sci. Tot. Environ. 375:297 - 309.

Zucker, L. and L. Brown (eds) Ag Drainage: Water Quality Impacts and Subsurface Drainage Studies in the Midwest, Ohio State University Extension Bulletin #871.

6

Cyclic Irrigation for Reducing Nutrients and Suspended Solids Loadings from Paddy Fields in Japan

Takehide Hama

Graduate school of Agriculture, Kyoto University
Japan

1. Introduction

The reduction of pollutants such as nitrogen, phosphorus, organic matter, and suspended solids discharged from non-point sources is an important aspect of improving water quality of downstream water areas (Reinelt et al., 1992; Gunes, 2008; Collins et al., 2010). Paddy fields, which produce rice as staple food in many countries, especially in the Asian monsoon region, and use large amounts of water during the rice growing season, are a major non-point source of pollution. Various environmental measures to reduce effluent load such as the reduction of chemical fertilizer (field-scale practices) (e.g., Choi & Nelson, 1996; Fan & Li, 2010) and reuse of drainage water (district-scale practices), are applied in paddy-field districts.

Cyclic irrigation (reuse of drainage water as irrigation water) is considered an effective water management practice for saving irrigation water resources and reducing effluent load from a paddy-field district. Cyclic irrigation was originally developed as a method for saving water in low-lying paddy fields (Kudo et al., 1995; Takeda et al., 1997) or terraced paddy fields (Tabuchi, 1986; Nakamura et al., 1998), where a stable and sufficient water source was not available. In a cyclic irrigation system, drainage water discharged from the paddy field is partially reused as irrigation water, so that the actual downstream effluent volume is decreased. Cyclic irrigation is also expected to decrease pollutant loads both because less water leaves the district and because some of the pollutants in the drainage water will be returned to the paddy field. Kubota et al. (1979) reported that cyclic irrigation with a recycling ratio (the ratio of reused water to drainage water) of 34% reduced nitrogen loads by 29% and phosphorus loads by 37%. In addition, cyclic irrigation system may increase the hydraulic retention time of nutrients in the paddy field and thereby enhance the purity of water leaving the field (Takeda et al., 1997; Feng et al., 2004, 2005; Takeda & Fukushima, 2006).

It has been also reported that the ability of cyclic irrigation to reduce loads of nutrients is directly proportional to the amount of reused water (Kaneki et al., 2003) and the recycling ratio (Hasegawa et al., 1982; Shiratani et al., 2004; Hitomi et al., 2006). However, the cyclic irrigation ratio, that is defined as the ratio of reused water to irrigation water, in paddy-field districts that have upstream areas is limited to low values due to large amount of uncontrollable inflow of water to the districts. Especially in paddy-field districts that capture industrial or domestic wastewater from upstream areas, irrigation water must have a large fresh water component to reduce the risks posed by pollutants including pathogens and heavy metals (Kaneki, 1989; Zulu et al., 1996).

Little is known about the ability of cyclic irrigation conducted with high recycling ratios to reduce loads from paddy-field districts. Furthermore, there have been few studies of this reduction effect as a function of the suspended solids load, even though suspended solids can cause various deleterious impacts (Bilotta & Brazier, 2008). In this chapter, we aimed to clarify the effects of cyclic irrigation with high cyclic irrigation ratio on water balance and nutrient and suspended solids loads in a paddy field and in the paddy-field district. We discussed the ability of cyclic irrigation to reduce the net exports of nutrient and suspended solids.

2. Description of study site

The study site was a low-lying paddy-field district located in Konohama district, on the southeastern edge of Lake Biwa (35°05′ N, 135°56′ E; Fig. 1). Lake Biwa is the largest lake in Japan and the most important water resource for the Kinki region, which includes Osaka and Kyoto. The mean annual temperature and rainfall are about 15 °C and 1550 mm (Japan Meteorological Agency, 2010). The district covers an area of about 1.5 km², of which more than 90% is used as paddy fields. Rotation crops are grown in about one-third of the paddy area each year on a 3-year cycle (Fig. 1). In rotation years, two rotation crops are grown, wheat and soybeans. The sequence of farming activities in paddy cultivation and rotation crop cultivation are summarized in Table 1. The rotation cropping cycle extends for one year, beginning in November with the sowing of wheat, which follows the harvesting of rice in September. The wheat is harvested in the middle of the next June. A crop of soybeans is sown soon after the harvesting of wheat and harvested in late November. The area is then left fallow and re-planted to paddy rice the following April. Chemical fertilizer (e.g., ammonium sulfate and calcium superphosphate) was not applied to soybeans. Base fertilizer was not applied to paddy fields after crop rotation.

The drainage and irrigation canals in the district are separated. There is no inflow of industrial or domestic wastewater from outside the study area into the drainage and irrigation canals. The drainage system contains lateral drainage canals, a main drainage canal, which passes through the district from north to south, and floodgates at both ends of the main drainage canal (Fig. 1). Rainfall runoff from the paddy fields and surplus irrigation water from the irrigation canals flow into the main drainage canal via the lateral drainage canals. All outflow of drainage water from the district is controlled by operation of the floodgates.

Two types of irrigation are practiced in the district: lake water irrigation and cyclic irrigation. In lake water irrigation, water is pumped from Lake Biwa into the irrigation canals. Under cyclic irrigation, drainage water in the main drainage canal is pumped into the irrigation canals and reused as irrigation water and water flowed from the lake to the drainage canal through the floodgate when the water level of drainage water decreased by evapotranspiration. There are two pump stations, one at the northern end and one at the southern end of the main drainage canal. The fields are not irrigated during the growing of rotation crops (i.e., crops other than paddy rice). The irrigation period is about 4 months, including a mid-summer drainage season of about 10 days. Cyclic irrigation is used from the beginning of the irrigation period to the mid-summer drainage season (referred to as the cyclic irrigation period), then lake water irrigation is used until the end of the irrigation period (the lake water irrigation period). The period from the end of the irrigation period to the beginning of the next irrigation period is referred to as the non-irrigation period.

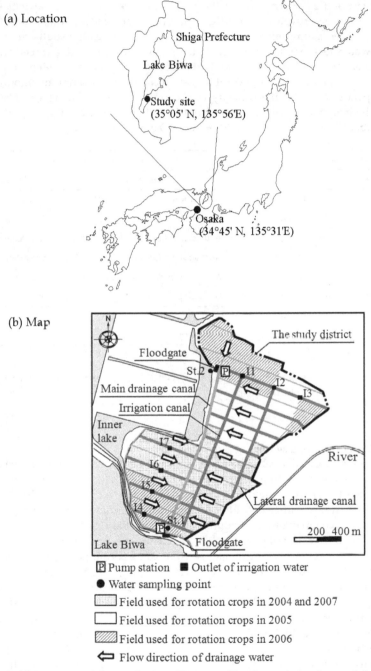

Fig. 1. (a) Location of the study site. (b) Map of land use, irrigation and drainage canals and of the water sampling points at the study site.

Pumps at the northern and southern ends of the main drainage canal have capacities of about 40 and 6 m³ min⁻¹, respectively. The northern pump station has two water inlets that connect to the lake and the main drainage canal, respectively, whereas the southern pump station has a single water inlet that only connects to the main drainage canal. Pumped water is delivered to outlets (points I1 to I7 in Fig. 1) through underground pipelines, and is supplied to the paddy fields through the several irrigation canals. The maximum amount of irrigation water depends solely on the capacity of the pumps, because there is no other source of water to the irrigation canals. Rainfall is not included in the irrigation water. The pumps operate for about 12 h per day, from 6:00 am to 6:00 pm.

Paddy rice		Rotation crops (wheat and soybeans)	
Farming activity	Timing	Farming activity	Timing
Base fertilization (N = 30, P = 30)	late April	Base fertilization (N = 60, P = 80)	November
Start of irrigation	late April	Sowing of wheat	November
Puddling, Sowing	late April – May	Additional fertilization (N = 30, P = 30)	late December
Additional fertilizaion (N = 10, P = 0)	late June	Additional fertilization (N = 30, P = 30)	late February
Mid-summer drainage	late June – July	Additional fertilization (N = 20, P = 0)	late April
Additional fertilizaion (N = 50, P = 0)	mid July	Harvesting	June
End of irrigation	late August	Sowing of soybeans	June
Harvesting	September	Harvesting	November

[a] N, the amouts of ferilizer of nitrogen (kgN ha⁻¹); P, the amouts of ferilizer of phosphorus (kgP ha⁻¹).

Table 1. The sequence of farming activities in the paddy-field district during the cultivation of paddy rice and cultivation of rotation crops.

3. Methodology

3.1 Field investigation of the paddy field

From 2004 to 2007, we performed weekly hydrological and water-quality measurement at two paddy fields in the district during the irrigation period each April to September. A location in the southwestern part of the district was used. The area of each paddy field is about 30 m × 100 m. The study fields were cultivated in normal farming methods for paddy rice, as other paddy fields in Japan or other countries (e.g., Liu et al., 2001; Kim et al., 2006). The paddy fields in the district were surrounded by earthen levees and ponded during the irrigation period except the mid-summer drainage season. Soil puddling is accompanied by tillage of the paddy fields to soften the soil before rice seedlings are transplanted at the beginning of the irrigation period. Nutrient and suspended solids concentration in a paddy field is especially high during the soil puddling season (Kaneki, 2003; Somura et al., 2009).

Figure 2 illustrates the components of water balance in the paddy field. Hydrological measurement instruments for rainfall (RT-5E, Ikeda-Keiki, Tokyo, Japan), air temperature (CS215L, Campbell Scientific, Inc., Logan, UT USA), wind velocity (014A-L, Campbell Scientific, Inc.), relative humidity (CS215L, Campbell Scientific, Inc.), and solar radiation (LP02-L, Campbell Scientific, Inc.) were installed in an open area at the southern pump station. Evapotranspiration was estimated by the Penman method (Penman, 1948) using data measured at the southern pump station and crop coefficient value for rice (Sakuratani & Horie, 1985). We measured the irrigation and runoff water (outflow through the outlet) flow rates delivered to and drained from the paddy fields using a Parshall flume set at the inlet and a triangular weir set at the outlet. A water-level meter (WT-HR, Intech Instruments Ltd., Christchurch, New Zealand) was set in each paddy field to calculate the change in water storage. The sum of percolation water volumes, which includes leakage water (lateral seepage to the drainage canal through the levee), was estimated from water balance calculations. Water balance in the paddy field is given by the following equation:

$$\Delta S = (R + I) - (ET + D + P) \tag{1}$$

where ΔS is the change in water storage, R is rainfall, I is irrigation water, ET is evapotranspiration, D is runoff water drained through the outlet, and P is percolation (all expressed in mm). However, percolation from the paddy field to the groundwater seems to be negligible because the district is low-lying and close to the lake and the groundwater level is high. Horizontal flow from or to the adjacent paddy fields is not considered since the district is located in the low-lying area.

The sum of rainfall and irrigation water minus runoff water $(R + I - D)$ was used to estimate the potential water demand of the paddy field in the district.

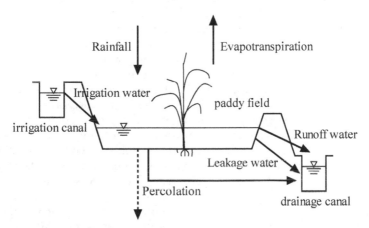

Fig. 2. Schematic diagram of water balance in the paddy field. Arrows indicate flow direction.

Each week, ponded water was sampled at the outlet of each paddy field and irrigation water was sampled at the outlet of the irrigation pipeline from the northern pump station (I1; Fig. 1). A small plastic tank was set near the northern pump station to collect rainfall water, which was sampled during the weekly field investigation. The manually sampled water was analyzed for total nitrogen (TN), dissolved total nitrogen (DTN), total phosphorous (TP), ammonium nitrogen (NH$_4$-N), nitrate nitrogen (NO$_3$-N), nitrite nitrogen (NO$_2$-N), and phosphate phosphorus (PO$_4$-P). Rainfall water sampled by the rain tank was analyzed for TN and TP.

Daily inputs and exports of nutrients in water were estimated by multiplying water nutrient concentrations by flow volumes. Percolation (including leakage) loss of nitrogen was estimated by using the TN in the ponded water as an estimate of nitrogen concentration in percolation or leakage water. Percolation loss of phosphorus was not estimated because it was thought phosphorus in the ponded water was strongly adsorbed to the paddy soil. Averaged data of the fields measurements in the fields were used for the estimation.

3.2 Field investigation of the study district

From 2006 to 2007, we performed weekly hydrological and water-quality measurement for the district during the irrigation period each April to September. Figure 3 is a conceptual diagram for water flow in the district. The flow rates of discharged drainage water during lake water irrigation or on rainy days and inflowing lake water during cyclic irrigation were measured using flow meters (2150 Area Velocity Flow Module, Teledyne Isco Inc., Lincoln, NE USA) installed at both ends of the main drainage canal. We estimated the volume of pumped water by multiplying the operating duration of the pumps by their capacity. We did not measure subsurface percolation from the district and assumed it to be negligible because the district is adjacent to the lake and the groundwater level is high, as mentioned above.

The characteristics of cyclic irrigation can be described by two different parameters (Kudo et al., 1995). One parameter is the ratio of reused water to pumped water (reused water plus lake water intake). Here, we refer to this parameter (α_{CI}) as the cyclic irrigation ratio. The other is the ratio of reused water to potential drainage water (reused water plus district drainage water discharged from the district), which is referred to as the recycling ratio and has often been used in previous studies (e.g., Kubota et al., 1979; Hasegawa et al., 1982; Hitomi et al., 2006). The recycling ratio depends more on drainage water than on reused water; in other words, the recycling ratio is affected more by water management in the paddy field and by weather conditions than is the cyclic irrigation ratio. For example, an increase in irrigation water into the paddy fields leads to a decrease in drainage water discharged from the district and results in a larger recycling ratio. Alternatively, in the case of cyclic irrigation after a rainfall event, increases in drainage water discharged from the district decrease the recycling ratio. Because of these problems with the recycling ratio, we have only analyzed and discussed the cyclic irrigation ratio. The mean cyclic irrigation ratio of the weekly measurements during the cyclic irrigation periods was 88% in 2006 and 82% in 2007, as described later.

The amount of surplus irrigation water can be approximately estimated as the volume of pumped water minus the volume of irrigation water used in the rice paddy fields (the percentage of the rice paddy fields in the district was set 66% in each investigation year). We defined the surplus irrigation water ratio (α_{SW}) as the ratio of surplus irrigation water to pumped water.

Water quality was measured within the district at weekly intervals from 2006 to 2007 by taking samples of drainage water at the southern end of the main drainage canal (St. 1; Fig. 1), irrigation water at the outlet of the pump (I1), and inner lake water (St. 2). In addition, an automatic water sampler (3700 Full-Size Portable Sampler, Teledyne Isco Inc.) was installed at St. 1 and used to sample drainage water daily at noon. Turbidimeters (Compact-CLW, JFE Alec Co., Ltd., Kobe, Japan) were set at both ends of the main drainage canal, set to a measurement interval of 20 min. The manually sampled water was analyzed for suspended solids (SS), TN, DTN, NH_4-N, NO_3-N, NO_2-N, TP and PO_4-P. Drainage water samples from the automatic sampler was analyzed for TN and TP.

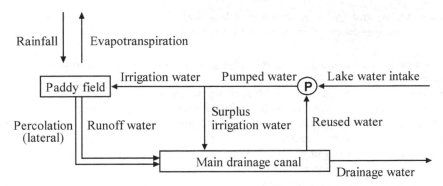

Fig. 3. Conceptual diagram of water flow in the study district. Upper-case "P" represents a pump and arrows indicate flow direction.

3.3 Water quality measurement

Nutrient concentration measurements were made using the following methods: TN and DTN were measured using an ultraviolet spectrophotometer (UV-1200, Shimadzu Corp., Kyoto, Japan) after alkaline potassium-peroxydisulfate digestion; TP by the molybdenum blue method after potassium-peroxydisulfate digestion; NH_4-N by the indo-phenol blue method; nitrate nitrogen by ion chromatography (LC-10A, Shimadzu Corp.); NO_2-N by the N-(1-naphthyl) ethylenediamine method; and PO_4-P by the molybdenum blue method. For this study, we defined SS as suspended matter with particle sizes ranging from 1 μm to 2 mm. Particulate-state and dissolved-state nutrients were also distinguished by filtering the sample with a 1-μm filter prior to analysis.

We calculated total inorganic nitrogen as the sum of NH_4-N, NO_3-N, and NO_2-N, and used PO_4-P as an estimate of total inorganic phosphorus. We calculated the total concentration of organic nitrogen or organic phosphorus as the difference between the total concentration and the total inorganic concentration. We calculated the concentration of particulate organic nitrogen as the difference between total nitrogen and dissolved total nitrogen, and the concentration of dissolved organic nitrogen as the difference between total dissolved nitrogen and total inorganic nitrogen.

Turbidimeter measurements were calibrated to convert turbidity readings to suspended solids content: calibration was performed by developing a relationship between field-measured turbidity and laboratory-measured suspended solids concentration of drainage water samples taken concurrently with turbidimeter readings.

3.4. Effects of cyclic irrigation on net exports of nutrients and suspended solids

The nutrient or SS loads are the product of the concentration and the water flow volume. Thus, the net export of nutrients or SS, L_{net} (kg ha^{-1} d^{-1}), is given by the following equation:

$$L_{net} = C_{out} Q_{out} - C_{in} Q_{in} \qquad (2)$$

where C is the concentration (mg L^{-1}), Q is the water flow volume (mm d^{-1}), and the subscripts *out* and *in* refer to outflow from and inflow into the district, respectively. In this case, C_{out} is the nutrient or SS concentration in the drainage water, Q_{out} is the amount of drainage water discharged from the district per day, C_{in} is the nutrient or SS concentration in the lake water, and Q_{in} is the amount of lake water intake per day. We estimated the relationship between the cyclic irrigation ratio (α_{CI}) and each variable.

3.4.1 Relationship between the cyclic irrigation ratio and the flow volume and the concentrations

The nutrient and SS concentration in the drainage water (C_{out}) during the normal irrigation periods may be proportional to the cyclic irrigation ratio because more pumping of drainage water leads to higher water flow and more erosion of bottom sediments in the main drainage canal. On the other hand, it is clear that C_{in} is essentially independent of the cyclic irrigation ratio because the impact of drainage water discharged from the district on the nutrients and SS concentration in the lake water would be negligible.

Consider the water flow during the cyclic irrigation period on a sunny day. Q_p represents the volume of pumped water and is about 20 mm d⁻¹. On sunny days, Q_p is the only driving force for water flow in the study district, which has a closed irrigation canal. We have assumed that water in the paddy field on a sunny day is mainly lost by evapotranspiration and that the amount of percolation or leakage water is negligible. In addition, runoff water occurs mainly during rainfall events. Thus, runoff and percolation (water flows from the paddy field into the main drainage canal via the lateral drainage canals) are not depicted in Fig. 4.

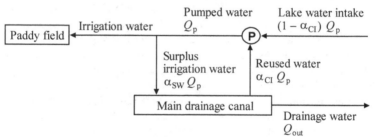

Fig. 4. Conceptual diagram of water flows under cyclic irrigation. Upper-case "P" represents a pump and arrows indicate flow direction.

Drainage water discharged from the district may potentially equal to the surplus irrigation water, $\alpha_{SW} Q_p$. Cyclic irrigation reduces the outflow of this potential drainage water due to reuse, $\alpha_{CI} Q_p$. Therefore, Q_{out} (actual drainage water) is written:

$$Q_{out} = (\alpha_{SW} - \alpha_{CI}) Q_p \tag{3}$$

The model of water flow illustrated in Fig. 4 does not consider temporary deficits of inflow water, which in practice are compensated for by decreases in drainage water flow in the main drainage canal. Equation (3) means that the upper limit of α_{CI} is α_{SW} when water flows out ($Q_{out} > 0$). If $\alpha_{SW} < \alpha_{CI}$ in Equation (3), another inflow of water from the lake must occur (negative Q_{out} in Fig. 4). In that case, $L_{net} = - (1 - \alpha_{SW}) C_{in} Q_{in}$; that is, under these conditions, L_{net} varies with α_{SW} and is negative for any α_{CI}.

Cyclic irrigation also reduces the inflow of water (lake water intake), Q_{in}, due to reuse. Thus, Q_{in} is written as follows:

$$Q_{in} = (1 - \alpha_{CI}) Q_p \tag{4}$$

These two parameters, α_{CI} and α_{SW}, can be taken as a supply- (source-) and demand- (user-) side water use parameter, respectively.

3.4.2 The effect of cyclic irrigation as a function of the cyclic irrigation ratio

Whether L_{net} is greater or less than zero indicates whether the effect of cyclic irrigation as a function of α_{CI} represents net contamination (cyclic irrigation increases the loadings from the district) or net purification (cyclic irrigation decreases the loadings). The neutral effect, $L_{net} = 0$, can be converted into the following equation by substituting the relationships between α_{CI} and Q_{out} (Equation (3)) and Q_{in} (Equation (4)) into Equation (2):

$$\frac{C_{out}}{C_{in}} = \frac{1-\alpha_{CI}}{\alpha_{SW} - \alpha_{CI}} \tag{5}$$

The effect of cyclic irrigation on L_{net} for a given α_{SW} value is illustrated in Fig. 5. If we replace the right side of Equation (5) with β, then β varies as a function of both α_{CI} and α_{SW}. Whether the effect of cyclic irrigation represents net contamination or net purification depends on whether the actual concentration ratio (C_{out}/C_{in}) for a given α_{CI} is above or below the β curve. In addition, the effect of cyclic irrigation at any α_{CI} is net purification if the concentration ratio is less than 1, because the value of β for any combination of α_{CI} and α_{SW} is greater than or equal to 1.

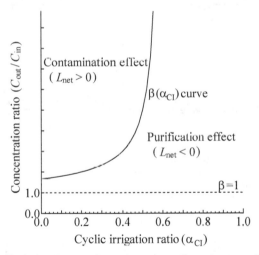

Fig. 5. The effect of cyclic irrigation on the net exports of nutrients and suspended solids (L_{net}) as a function of the cyclic irrigation ratio (α_{CI}).

4. Results and discussion

4.1 Characteristics of water balance and nutrient loads in a paddy field
4.1.1 Water balance in the paddy field

Figure 6 shows daily variations in inflow water (rainfall and irrigation water) and outflow water (evapotranspiration and runoff water) in the paddy field during each irrigation period from 2004 through 2007.

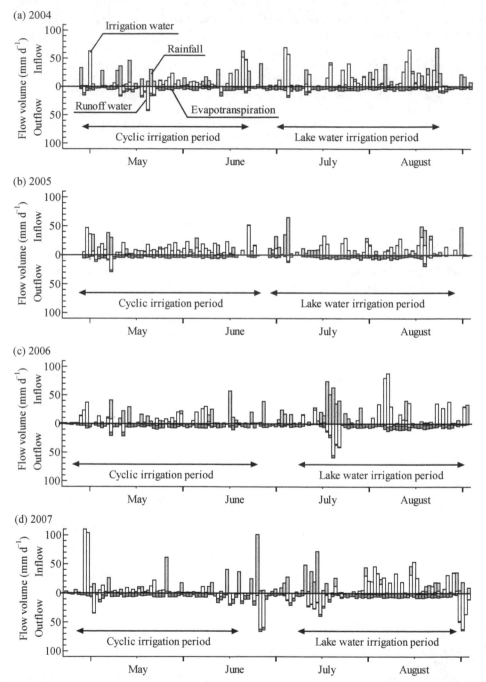

Fig. 6. Daily irrigation water, rainfall, runoff water, and evapotranspiration in the paddy fields during the irrigation period in (a) 2004, in (b) 2005, in (c) 2006 and in (d) 2007.

A complementary relationship was seen between irrigation water and rainfall because the pumps were not operated on rainy days. Runoff mainly occurred during and after rainfall. Runoff of 34 mm d^{-1} in early May in 2007 occurred because of artificial drainage by the farmer, who intended to dry the paddy fields and then transplant rice earlier. In addition, the large amount of runoff water during the 2007 mid-summer drainage season was due both to rainfall and to the temporary removal of shuttering boards at the outlets of the paddy fields during the irrigation season.

Table 2 shows the water balances for the paddy field during the irrigation periods in 2004–2007. The water level of the field at the begging and the end of each period was 0 (i.e., $\Delta S = 0$). Total amounts of rainfall during the irrigation periods ranged from 469 mm in 2005 to 779 mm in 2006. Weather conditions in the investigation years except 2005 were considered normal because the amounts of rainfall during the irrigation period were within the range of the mean ± standard deviation (766 ± 205 mm) from three decades of data (1980–2009) in Otsu City, which is near the study district (Japan Meteorological Agency, 2010).

Year	Period[a]	Inflow (mm)		Outflow (mm)	
		Rainfall	Irrigation water	Evapotranspiration	Runoff water
2004	CI period (28 April – 20 June)	313	377	213	178
	Mid-summer drainage season[b]	73	0	37	8
	LWI period (1 July – 21 August)	139	624	252	65
	Total	525	1001	502	251
2005	CI period (25 April – 30 June)	210	563	210	71
	Mid-summer drainage season	115	0	7	14
	LWI period (6 July – 28 August)	139	379	179	46
	Total	464	942	396	131
2006	CI period (24 April – 25 June)	277	275	212	72
	Mid-summer drainage season	102	0	37	8
	LWI period (8 July – 31 August)	400	500	241	213
	Total	779	775	490	293
2007	CI period (25 April – 23 June)	281	356	273	138
	Mid-summer drainage season	175	0	38	157
	LWI period (6 July – 28 August)	319	506	261	193
	Total	775	862	572	488

[a] CI, cyclic irrigation; LWI, lake water irrigation.

[b] From the end of CI period to the beginning of LWI period.

Table 2. Mean water balances for the paddy fields during the irrigation period.

The total amount of irrigation water each year was inversely proportional to rainfall, and the sum of rainfall and irrigation water during each irrigation period was fairly constant (about

1400–1600 mm each year). About 1100–1200 mm was estimated as the potential water demand during the irrigation period (without the mid-summer drainage season). The difference between total inflow and total outflow during the irrigation period, which equals the sum of stored water and percolation, ranged from 577 mm (in 2007) to 879 mm (in 2005). From these results, it was estimated that water loss from the paddy fields by percolation was about 7 mm d^{-1} (at most) during the irrigation period. The amount of water lost through percolation was likely more than that lost through evapotranspiration, which ranged from 396 mm (in 2005) to 572 mm (in 2007).

The reason that the difference in water management practices in 2004–2007 was not reflected in the water balances of each irrigation period is because irrigation water was supplied only from the pumps, and the irrigation schedule for each field depended on the pump operation. In other words, supply-side water management practices seemed to have a greater influence on water balance in the paddy fields than did individual farmers' management practices. An irrigation system with a closed irrigation canal (receiving no inflow of water from outside the area) that enables the paddy-field district to conduct cyclic irrigation with a high cyclic irrigation ratio, combined with supply-side water management (e.g., stopping the pumps during rainfall events), can provide efficient use of rainfall for crop irrigation, though such an irrigation system is less flexible for meeting the water use demands of individual farmers.

Total amounts of pumped water in the irrigation periods were 1528 mm in 2004, 1720 mm in 2005, 1737 mm in 2006 and 1681 mm in 2007, and the amounts of surplus irrigation water (= the volume of pumped water minus the volume of irrigation water used in the rice paddy fields) were therefore 867 mm (= 1528 mm – 0.66 × 1001 mm) in 2004, 1098 mm in 2005, 1226 mm in 2006 and 1112 mm in 2007. The overall surplus irrigation water ratio in the district in the irrigation periods was 57% (= 867 mm / 1528 mm × 100) in 2004, 64% in 2005, 71% in 2006 and 66% in 2007.

4.1.2 Nutrient loads in the paddy field

Figure 7 shows the temporal variations in TN and TP of irrigation water and ponded water during the 2007 irrigation period. Nutrient concentrations in the ponded water were higher than in irrigation water during the puddling season. In contrast, nutrient concentrations in the ponded water were similar to those in irrigation water during the irrigation period following the puddling season (i.e., the normal irrigation period referred to in this paper). These results indicate that the quality of irrigation water has a large influence on ponded water during the normal irrigation period.

Nutrient concentrations in irrigation water in 2004–2007 are shown in Table 3. The trends for each nutrient component in irrigation water were similar over the study years, except for lower nutrient concentrations of TN and TP during the puddling season in 2004, which may have been caused by dilution in successive rainfall events during that season. Nutrient concentrations in irrigation water in the study years were highest during the puddling season (TN = 3.26–4.07 mg L^{-1}, TP = 0.04–0.29 mg L^{-1}), and higher during the cyclic irrigation period (TN = 1.76–2.27 mg L^{-1}, TP = 0.09–0.24 mg L^{-1}) than during the lake water irrigation period (TN = 0.53–0.73 mg L^{-1}, TP = 0.04–0.06 mg L^{-1}). The high nutrient concentrations in irrigation water during the puddling season are likely due to dissolution and leaching of nutrients from paddy soil.

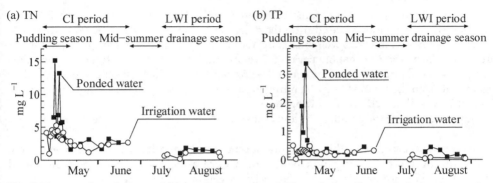

Fig. 7. Temporal variations of (a) total nitrogen and (b) total phosphorus in irrigation water and ponded water during the irrigation period in 2007. CI, cyclic irrigation; LWI, lake water irrigation.

Year	Period[a]	Water quality (mg L^{-1})[d]							
		TN	DTN	NH$_4$-N	NO$_3$-N	NO$_2$-N	TP	PO$_4$-P	n[f]
2004	Puddling season[b]	3.26	—[e]	0.35	1.98	0.05	0.04	0.03	4
	CI period[c]	1.76	—	0.26	0.44	0.02	0.09	0.05	5
	LWI period	0.54	—	0.06	0.05	0.00	0.04	0.01	5
2005	Puddling season	4.07	2.35	0.43	1.03	0.00	0.24	0.04	4
	CI period	2.04	1.13	0.26	0.21	0.00	0.18	0.02	6
	LWI period	0.73	0.52	0.04	0.04	0.00	0.03	0.01	3
2006	Puddling season	3.91	1.92	0.15	0.72	0.00	0.26	0.04	5
	CI period	1.83	0.85	0.17	0.41	0.00	0.16	0.03	4
	LWI period	0.53	0.40	0.05	0.05	0.00	0.06	0.02	5
2007	Puddling season	4.00	2.81	0.46	1.77	0.00	0.29	0.03	10
	CI period	2.27	1.41	0.40	0.36	0.00	0.24	0.02	8
	LWI period	0.72	0.53	0.05	0.04	0.00	0.05	0.01	7

[a] CI, cyclic irrigation; LWI, lake water irrigation.

[b] From the beginning of the irrigation period to early May.

[c] Cyclic irrigation period after the puddling season.

[d] TN; total nitrogen; DTN, dissolved total nitrogen; TP, total phosphorus.

[e] No data.

[f] The number of sapmles.

Table 3. Mean nutrient concentrations in irrigation water in 2004–2007.

The mean concentration of inorganic nitrogen (= NH_4-N + NO_3-N + NO_2-N) in cyclic irrigation water (irrigation water during the cyclic irrigation period) in 2004–2007 was 0.47–0.76 mg L^{-1}, whereas the level was constantly about 0.1 mg L^{-1} in lake water (irrigation water during the lake water irrigation period). Organic nitrogen (= total nitrogen – inorganic nitrogen) in cyclic irrigation water was 1.04–1.57 mg L^{-1}, of which 60–80% was particulate organic nitrogen. In contrast, organic nitrogen in lake water was 0.43–0.65 mg L^{-1}, of which about 70% was dissolved organic nitrogen.

The inputs and exports of nitrogen and phosphorus are shown in Table 4 and Table 5 respectively.

In each of the four years of this study, the inputs of nitrogen from irrigation water were greater during the cyclic irrigation period than during the lake water irrigation period (Table 4). Exports of nitrogen during the cyclic irrigation period were also larger than that during the lake water irrigation period. Percolation (including leakage) loss of nitrogen was estimated as from 7.2 kg ha^{-1} (in 2006) to 12.3 kg ha^{-1} (in 2005). However, it seems that the actual percolation loss of nitrogen was less than the estimated values, which might be because nutrients in the ponded water were mainly in an organic state and easily adsorbed by the paddy soil as water flowed through.

Year	Period[a]	Inputs (kg ha^{-1})		Exports (kg ha^{-1})		
		Rainfall	Irrigation water	Runoff water	Percolation	Net[c]
2004	CI period (28 April – 20 June)	1.8	7.2	6.7	5.3	3.0
	Mid-summer drainage season[b]	0.7	0.0	0.1	0.0	-0.6
	LWI period (1 July – 21 August)	1.4	2.8	0.9	2.4	-0.9
2005	CI period (25 April – 30 June)	1.6	10.0	2.3	10.1	-0.8
	Mid-summer drainage season	0.9	0.0	0.1	0.0	-0.8
	LWI period (6 July – 28 August)	1.0	3.4	0.8	2.2	-1.4
2006	CI period (24 April – 25 June)	1.4	7.1	2.2	4.9	-1.4
	Mid-summer drainage season	0.8	0.0	0.3	0.0	-0.5
	LWI period (8 July – 31 August)	3.9	3.5	2.0	2.3	-3.1
2007	CI period (25 April – 23 June)	2.2	12.5	5.0	5.1	-4.6
	Mid-summer drainage season	1.4	0.0	3.4	0.0	2.0
	LWI period (6 July – 28 August)	2.5	4.6	2.9	2.7	-1.5

[a] CI, cyclic irrigation; LWI, lake water irrigation.

[b] From the end of CI period to the beginning of LWI period.

[c] Runoff water + Percolation −(Rainfall + Irrigation water).

Table 4. Nitrogen loads in the paddy fields during the irrigation period.

Net exports of nitrogen from a paddy field, which is estimated as exports (= runoff water and percolation water) minus inputs (= rainfall and irrigation water), indicates whether the water management practices associated with that field may increase or decrease the nitrogen load. A negative value of net exports means that the paddy field decreased nitrogen load during the calculation period. In this study, net exports of nitrogen were negative during all lake water irrigation periods. A similar situation was observed for a paddy field adjacent to Kasumigaura Lake (the second largest lake in Japan), a region that is aiming to remove nitrogen from river water (Zhou & Hosomi, 2008). Our data indicate that lake water irrigation may remove nitrogen from the outside water area (i.e., Lake Biwa), whereas cyclic irrigation using a high cyclic irrigation ratio probably does not because almost all the nitrogen in cyclic irrigation water was originally input as fertilizer. In this case, the major benefit of cyclic irrigation is considered to be the return of nitrogen to the paddy field, which possibly leads to a reduction in fertilizer usage. From other viewpoints, it may be said that cyclic irrigation system realizes the smallest nitrogen cycle, with the paddy field acting as a means of self-purification in the district.

Year	Period[a]	Inputs (kg ha^{-1})		Exports (kg ha^{-1})	
		Rainfall	Irrigation water	Runoff water	Net[c]
2004	CI period (28 April – 20 June)	0.04	0.50	1.18	0.64
	Mid-summer drainage season[b]	0.02	0.00	0.02	0.00
	LWI period (1 July – 21 August)	0.03	0.20	0.10	-0.13
2005	CI period (25 April – 30 June)	0.04	0.70	0.46	-0.28
	Mid-summer drainage season	0.02	0.00	0.01	-0.01
	LWI period (6 July – 28 August)	0.02	0.30	0.10	-0.22
2006	CI period (24 April – 25 June)	0.05	0.46	0.24	-0.27
	Mid-summer drainage season	0.01	0.00	0.04	0.03
	LWI period (8 July – 31 August)	0.07	0.32	0.33	-0.06
2007	CI period (25 April – 23 June)	0.05	0.95	1.06	0.06
	Mid-summer drainage season	0.03	0.00	0.62	0.59
	LWI period (6 July – 28 August)	0.06	0.30	0.61	0.25

[a] CI, cyclic irrigation; LWI, lake water irrigation.

[b] From the end of CI period to the beginning of LWI period.

[c] Runoff water — (Rainfall + Irrigation water).

Table 5. Phosphorus loads in the paddy fields during the irrigation period.

Similar to inputs of nitrogen, inputs of phosphorus from irrigation water was larger during the cyclic irrigation period than during the lake water irrigation period (Table 5). The export of phosphorus, however, was relatively large, and the net exports were positive during the irrigation periods in both 2004 and 2007. The large exports of phosphorus during the cyclic irrigation periods in 2004 and 2007 were most likely due to rainfall and artificial drainage, respectively. The influence of weather conditions and water management appear to have a greater influence on exports of phosphorus than on exports of nitrogen. Therefore, water management practices at the paddy-field level (e.g., drying paddy fields without artificial drainage) are important for reducing the export of phosphorus, even though practices at the district level (e.g., conduction of cyclic irrigation throughout the entire irrigation period) can further reduce net export, as described next.

4.2 Characteristics of water balance and nutrient and suspended solids loads in the paddy-field district
4.2.1 Water balance in the study district

Daily variations in rainfall and drainage water from the district through the floodgates in 2006 and 2007 are shown in Fig. 8. Drainage water was not released during the cyclic irrigation periods, except during rainfall events, whereas during the lake water irrigation periods drainage water of more than 10 mm d^{-1} was released even on sunny days. The amount of drainage water discharged from the district on sunny days during the lake water irrigation periods nearly equaled the amount of surplus irrigation water, suggesting that cyclic irrigation reduced the outflow of surplus irrigation water from the district.

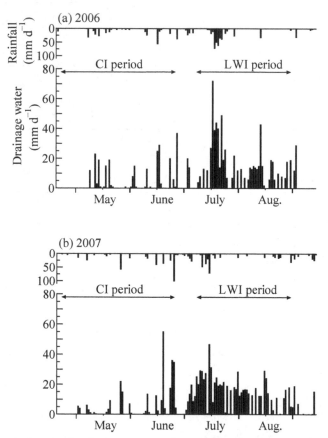

Fig. 8. Daily variations in the drainage water from the study district and rainfall during the irrigation in (a) 2006 and in (b) 2007: CI, cyclic irrigation; LWI, lake water irrigation.

Table 6 shows the water balances in the district during the irrigation periods.

Year	Period[a]	Inflow (mm)		Outflow (mm)	
		Rainfall	Lake water intake	Evapotranspiration	Drainage water
2006	CI period (24 April – 25 June)	277	134	186	221
	Mid-summer drainage season[b]	102	0	29	71
	LWI period (8 July – 31 August)	400	582	237	707
	Total	779	716	452	999
2007	CI period (25 April – 23 June)	281	174	248	237
	Mid-summer drainage season	175	0	31	94
	LWI period (6 July – 28 August)	319	669	258	768
	Total	775	843	537	1099

[a] CI, cyclic irrigation; LWI, lake water irrigation.

[b] From the end of CI period to the beginning of LWI period.

Table 6. Water balance in the study district during the irrigation periods in 2006 and 2007.

Although the amounts of pumped water during the cyclic irrigation periods (1111 mm in 2006 and 962 mm in 2007) were larger than those during the lake water irrigation periods (626 mm in 2006 and 719 mm in 2007), the amounts of lake water intake during the cyclic irrigation periods were less than those during the lake water irrigation periods, because pumped water was mainly supplied by the reuse of drainage water during cyclic irrigation. The smaller amounts of drainage water discharged from the district during the cyclic irrigation periods were also due to the reuse of drainage water. The amounts of reused water (pumped water minus lake water intake) during the cyclic irrigation periods were 977 mm in 2006 and 788 mm in 2007.

4.2.2 Nutrient and suspended solids concentrations in the drainage water

Temporal variations in nutrient and suspended solids concentrations during the irrigation periods in 2006 and in 2007 are shown in Fig. 9. The variation trends were similar in 2006 and 2007. The nutrient concentrations were higher during the puddling season and on days on which rain fell. Nutrient concentrations on fine days during the irrigation period ranged from 1.0 to 2.0 mg L^{-1} for TN and from 0.10 to 0.20 mg L^{-1} for TP. The nutrient concentrations in the drainage water were higher during the cyclic irrigation period than during the lake water irrigation period. The SS concentration was also high during the puddling season (from late April to mid-May) and during heavy rainfall events; the SS concentration was more than 100 mg L^{-1} at its peak during the puddling season. The SS concentration on sunny days during the cyclic irrigation periods after the puddling season was about 20 mg L^{-1} and was higher than about 10 mg L^{-1} on sunny days during the lake water irrigation periods. The nutrient and SS concentrations in irrigation water during the cyclic irrigation periods nearly equaled the nutrient and SS concentrations in the drainage water because the cyclic irrigation ratios during the cyclic irrigation periods were high and the dilution volumes from the lake water were small.

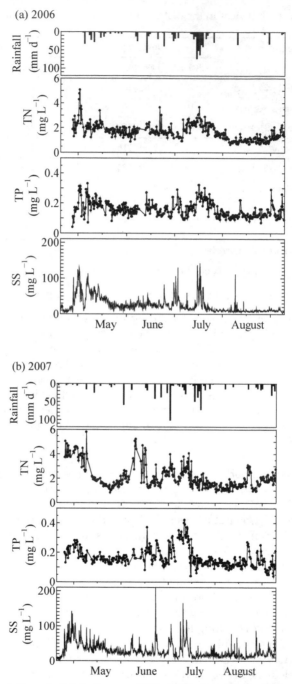

Fig. 9. Temporal variations in total nitrogen (TN), total phosphorus (TP) and suspended solids (SS) concentration in the drainage water in (a) 2006 and in (b) 2007.

4.2.3 Nutrient and suspended solids loads in the study district

The inputs of TN and TP in rainfall and lake water and exports of TN and TP in discharged drainage water during the irrigation periods are shown in Table 7 and Table 8. The total exports of nitrogen and phosphorus were about 20 kg ha[-1] and 2.0 kg ha[-1], respectively. Net exports of nitrogen and phosphorus from the paddy field (= runoff water – rainfall – lake water intake) were positive during all irrigation periods. Therefore, it is suggested that the study district acts as source of nutrients. The export of nutrients during the cyclic irrigation periods was less than that during the lake water irrigation period, in line with the small amounts of water discharged (Table 6). Some of the nutrient exports during the lake water irrigation periods were caused by the discharge of surplus irrigation water on fine days (Fig. 8), although nutrient concentrations in the drainage water were lower than during cyclic irrigation.

Year	Period[a]	Inputs (kg ha^{-1})		Exports (kg ha^{-1})	
		Rainfall	Lake water intake	Drainage water	Net[c]
2006	CI period (24 April – 25 June)	1.4	1.1	4.4	1.9
	Mid-summer drainage season[b]	0.8	0.0	1.1	0.3
	LWI period (8 July – 31 August)	3.9	4.5	12.9	4.5
2007	CI period (25 April – 23 June)	2.2	2.8	5.2	0.2
	Mid-summer drainage season	1.4	0.0	2.3	0.9
	LWI period (6 July – 28 August)	2.5	4.6	13.7	6.6

[a] CI, cyclic irrigation; LWI, lake water irrigation.

[b] From the end of CI period to the beginning of LWI period.

[c] Drainage water − (Rainfall + Lake water intake).

Table 7. Nitrogen loads in the study district.

Year	Period[a]	Inputs (kg ha^{-1})		Exports (kg ha^{-1})	
		Rainfall	Lake water intake	Drainage water	Net[c]
2006	CI period (24 April – 25 June)	0.00	0.10	0.40	0.30
	Mid-summer drainage season[b]	0.00	0.00	0.20	0.20
	LWI period (8 July – 31 August)	0.10	0.30	1.40	1.00
2007	CI period (25 April – 23 June)	0.10	0.20	0.40	0.10
	Mid-summer drainage season	0.00	0.00	0.20	0.20
	LWI period (6 July – 28 August)	0.10	0.40	1.40	0.90

[a] CI, cyclic irrigation; LWI, lake water irrigation.

[b] From the end of CI period to the beginning of LWI period.

[c] Drainage water − (Rainfall + Lake water intake).

Table 8. Phosphorus loads in the study district.

Table 9 shows the SS loads in the district during the irrigation periods. The exports of SS during the cyclic irrigation periods were less than those during the lake water irrigation periods, even though the cyclic irrigation periods included the puddling seasons, when SS concentration in runoff water from the paddy fields was very high. Clearly, the exports of SS from the district were reduced during the cyclic irrigation periods. Another effect of cyclic irrigation is to return SS to the paddy fields along with the reused water. The return of SS to the paddy field during cyclic irrigation, estimated from the product of the SS concentration and the amount of irrigation water, was 118 kg ha^{-1} in 2006 and 199 kg ha^{-1} in 2007.

Year	Period[a]	Inputs (kg ha^{-1})		Exports (kg ha^{-1})	
		Rainfall	Lake water intake	Drainage water	Net[c]
2006	CI period (24 April – 25 June)	0	7	90	83
	Mid-summer drainage season[b]	0	0	35	35
	LWI period (8 July – 31 August)	0	26	152	126
2007	CI period (25 April – 23 June)	0	28	80	52
	Mid-summer drainage season	0	0	39	39
	LWI period (6 July – 28 August)	0	30	183	153

[a] CI, cyclic irrigation; LWI, lake water irrigation.

[b] From the end of CI period to the beginning of LWI period.

[c] Drainage water − (Rainfall + Lake water intake).

Table 9. Suspended solids loads in the study district.

4.3 Effects of cyclic irrigation on net exports of nutrients and suspended solids

In this section, we discuss the effect of cyclic irrigation on reducing the net exports of nutrients and SS (Equation (2)) from the district on a sunny day during the normal irrigation period, which represents the irrigation period after the puddling season.

We plotted the relationship between the cyclic irrigation ratio (α_{CI}) and the nutrient and SS concentration in the drainage water (C_{out}) during the normal irrigation periods (Fig. 10).

C_{out} may be proportional to α_{CI}. The distribution of the fields under rotation crops (i.e., crops other than paddy rice) may also influence C_{out}. The fields were distributed around the northern and southern of the district in 2006 and around the center of the district in 2007. We hypothesize that more of the SS in rainfall runoff from the field under crop rotation settled out in the main drainage canal in 2007 than in 2006 because the distance from the rotation crop areas to the floodgates was shorter in 2006. Accordingly, the cyclic irrigation may have led to higher C_{out} on a sunny day in 2007 than in 2006. The mean TN concentrations (C_{in} for nitrogen) were 0.78 mg L^{-1} in 2006 and 0.68 mg L^{-1} in 2007. The mean TP concentrations (C_{in} for phosphorus) were 0.06 mg L^{-1} in 2006 and 2007. The mean SS concentrations (C_{in} for SS) were 4.5 mg L^{-1} in 2006 and 2007.

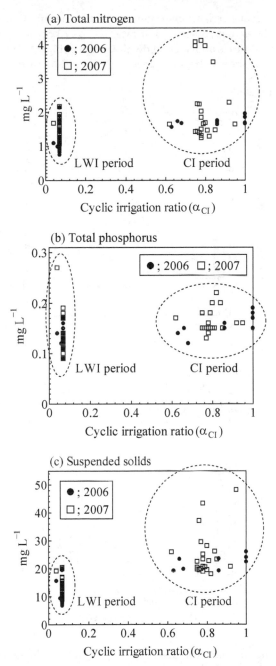

Fig. 10. Relationship between the cyclic irrigation ratio (α_{CI}) and (a) total nitrogen, (b) total phosphorus and (c) suspended solid concentration in the drainage water: CI, cyclic irrigation; LWI, lake water irrigation.

Whether the effect of cyclic irrigation represents net contamination or net purification depends on whether the actual concentration ratio (C_{out}/C_{in}) for a given α_{CI} and α_{SW} (the surplus irrigation water ratio) is above or below the β curve (Fig. 5). The β is calculated from Equation (5). Figure 11 shows the measured concentration ratios during the normal

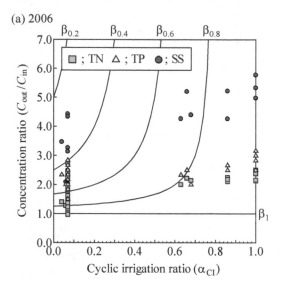

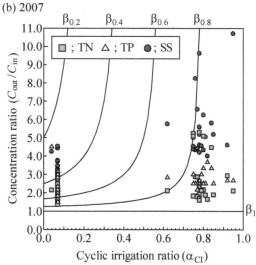

Fig. 11. Measured concentration ratios of total nitrogen (TN), total phosphorus (TP) and suspended solids (SS) in (a) 2006 and in (b) 2007. The subscript for each β value (= $[1 - \alpha_{CI}] / [\alpha_{SW} - \alpha_{CI}]$) represents the value of the surplus irrigation ratio (α_{SW}) used to calculate the β curve.

irrigation periods, as well as five β curves for various values of α_{SW} (=0.2, 0.4, 0.6, 0.8, and 1.0). It is clearly that the effect of cyclic irrigation at high α_{CI} will be net purification even if α_{SW} is high, whereas at low α_{CI} the effect of cyclic irrigation may be net contamination when α_{SW} is greater than 0.6. Though intermediate values of the cyclic irrigation ratio were not used in the district, Fig. 11 indicates that conducting cyclic irrigation with a moderate value of α_{CI} will not necessarily cause net purification if increasing α_{CI} increases the concentration ratio. The possibility that increasing α_{CI} increases C_{out} is shown in Fig. 10.

α_{SW} is another important parameter to consider when predicting the effect of cyclic irrigation. When the value of α_{SW} is high, the effect of cyclic irrigation is net contamination for almost all values of α_{CI}. In contrast, the effect of cyclic irrigation is net purification for almost all value of α_{CI} when α_{SW} has a low value. α_{SW} is strongly influenced by weather conditions, especially evapotranspirational demand and rainfall, and by water management practices in the paddy fields. In fact, daily α_{SW} ranged from 0.3 to 0.9 and was high in the spring and low in the summer in the study district.

Based on these results, two approaches can be used to produce net purification through cyclic irrigation; increasing α_{CI} and decreasing α_{SW}. Both parameters interact to determine the net effect of cyclic irrigation. Fig. 11 suggests that improving both parameters simultaneously will reduce net exports of nutrients and SS more effectively than improving either parameter alone.

Reduction of effluent loads in the drainage canals is also important, because the canals connect the fields with the downstream water bodies and function as a sink or source of nutrients and SS. However, there is little research on the dynamics of nutrients and SS in agricultural drainage canals: It is essential for the appropriate management of drainage canals to understand the deposition and resuspension of SS and the adsorption and dissolution of nutrients on sediment.

5. Conclusions

It is suggested that a cyclic irrigation system that enables the paddy-field district to use a high cyclic irrigation ratio may lead to more efficient use of rainfall for crop irrigation because there was a clear inverse relationship between amount of irrigation water applied and amount of rainfall each year. Drainage water discharged from the district may potentially equal to the surplus irrigation water on a sunny day during the normal irrigation period. Cyclic irrigation reduces the outflow of this potential drainage water due to reuse.

The export of nutrients from the district during the cyclic irrigation periods was less than that during the lake water irrigation period. It is also confirmed that cyclic irrigation can effectively reduce the suspended solids load during the puddling season when the suspended solids concentration in drainage water is high. The influence of weather conditions and water management appear to have a greater influence on exports of phosphorus than on that of nitrogen.

The effect of cyclic irrigation on the net nutrient and suspended solids exports can be represented by three ratios: the concentration ratio, which represents the ratio of the nutrient and suspended solids concentrations in drainage water to that in lake water; the cyclic irrigation ratio, which represents the ratio of the volume of reused water to that of pumped water in cyclic irrigation; and the surplus irrigation water ratio, which represents the ratio of the volume of surplus irrigation water to that of pumped water. Both the latter parameters interact to determine the net effect of cyclic irrigation. Simultaneously increasing the cyclic irrigation ratio and decreasing the surplus irrigation water ratio is important to maximize purification effect.

6. Acknowledgement

We thank the Konohama Land Improvement District, the Konohama Agricultural Union, and the Shiga Prefecture Office for providing access to the paddy fields for investigation and for providing daily reports on water management and farming activity in the paddy fields. The research described in this paper was partly funded by a grant from the Kinki Regional Agricultural Administration Office of the Japanese Ministry of Agriculture, Forestry and Fisheries, and by a Grant-in-Aid for Scientific Research from the Japan Society for the Promotion of Science.

7. Reference

Bilotta, G. S. & Brazier, R. E. (2008). Understanding the influence of suspended solids on water quality and aquatic biota. *Water Res.*, Vol.42, pp.2849–2861.

Choi, J. M. & Nelson, P. V. (1996). Developing a slow-release nitrogen fertilizer from organic sources: II. Using poultry feathers. *J. Am. Soc. Hort. Sci.*, Vol.121, No.4, pp.634–638.

Collins, K. A., Lawrence, T. J., Stander, E. K., Jontos, R. J., Kaushal, S. S., Newcomer, T. A., Grimm, N. B. & Cole Ekberg, M. L. (2010). Opportunities and challenges for managing nitrogen in urban stormwater: A review and systhesis. *Ecol. Eng.*, Vol.36, pp.1507–1519.

Gunes, K. (2008). Point and nonpoint sources of nutrients to lakes – ecotechnological measures and mitigation methodologies – case study. *Ecol. Eng.*, Vol.34, pp.116–126.

Fan, X. H. & Li, Y. C. (2010). Nitrogen release from slow-release fertilizers as affected by soil type and temperature. *Soil Sci. Soc. Am. J.*, Vol.74, pp.1635–1641.

Feng, Y. W., Yoshinaga, I., Shiratani, E., Hitomi, T. & Hasebe, H. (2005). Nutrient balance in a paddy field with a recycling irrigation system. *Water Sci. Technol.*, Vol.51, No.3, pp.151–157.

Feng, Y. W., Yoshinaga, I., Shiratani, E., Hitomi, T. & Hasebe H. (2004). Characteristics and behavior of nutrients in a paddy field area equipped with recycling irrigation system. *Agric. Water Manage.*, Vol.68, pp.47–60.

Hasegawa, K., Kobayashi, M., Nakamura, M. & Nakata, H. (1982). The effect of return flow irrigation on balance of water polluting material in paddy fields (1). *Report of Shiga Prefecture Agricultural Center*, Vol.24, pp.65–78 (in Japanese).

Hitomi, T., Yoshinaga, I., Feng, Y. W. & Shiratani, E. (2006). Nitrogen removal function of recycling irrigation system. *Water Sci. Tech.*, Vol.53, No.2, pp.101–109.

Kaneki, R. (1989). Pollutant reduction to Lake Biwa by using a cyclic irrigation system. *J. JSIDRE*, Vol.57, No.7, pp.39–44 (in Japanese).

Kaneki, R., Nakamura, M., Izumi, M. & Himeno, Y. (2003). Water purification by lagoon and cyclic irrigation. *J. JSIDRE*, Vol.71, No.9, pp.31–36 (in Japanese).

Kaneki, R. (2003). Reduction of effluent nitrogen and phosphorus from paddy fields. *Paddy Water Environ.*, Vol.1, pp.133–138.

Kim, J. S., Oh, S. Y. & Oh, K. Y. (2006). Nutrient runoff from a Korean rice paddy watershed during multiple storm events in the growing season. *J. Hydrol.*, Vol.327, pp.128-139.

Kubota, H., Tabuchi, T., Takamura, Y. & Suzuki, S. (1979). Water and material (N, P) balance in the paddy fields along Lake Kasumigaura. *Trans. JSIDRE*, Vol.84, pp.22–28 (in Japanese).

Kudo, A., Kawagoe, N. & Sasanabe, S. (1995). Characteristics of water management and outflow load from a paddy field in a return flow irrigation area. *J. JSIDRE*, Vol.63, No.2, pp. 49–54 (in Japanese).

Liu C.W., Chen, S.K., Jou, S.W. & Kuo, S.F. (2001). Estimation of the infiltration rate of a paddy field in Yun-Lin, Taiwan. *Agric. Systems*, Vol.68, pp.41-54.

Japan Meteorological Agency (July 2010). Automated Meteorological Data Acquisition System (AMeDAS), In: *Ostu City*, 01.07.2010, Available from http://www.jma.go.jp

Nakamura, Y., Manpuku, K. & Koga, Y. (1998). Conservation of water environment on cyclic irrigation system in the Inbanuma basin. *J. JSIDRE*, Vol.66, No.12, pp.37–44 (in Japanese).

Penman, H. L. (1948). Natural evaporation from open water, bare soil and grass. *Proc. Roy. Soc. London, A*, Vol.193, pp.120–145.

Reinelt, L. E., Horner, R. R. & Castensson, R. (1992). Non-point source water pollution management: improving decision-making information through water quality monitoring. *J. Environ. Manage.*, Vol.34, pp.15-30.

Sakuratani, T. & Horie, T. (1985). Studies on evapotranspiration from crops (1) on seasonal changes, vertical differences and the simplified methods of estimate in evapotranspiration of paddy rice. *J. Agric. Meteorol.*, Vol.41, No.1, pp.45–55 (in Japanese).

Shiratani, E., Yoshinaga, I., Feng, Y. W. & Hasebe, H. (2004). Scenario analysis for reduction of effluent load from an agricultural area by recycling the run-off water. *Water Sci. Tech.*, Vol.49, No.3, pp.55–62.

Somura, H., Takeda, I. & Mori, Y. (2009). Influence of puddling procedures on the quality of rice paddy drainage water. *Agric. Water Manage.*, Vol.96, pp.1052–1058.

Tabuchi, T. (1986). Nitrogen outflow diagram in a small agricultural area. *Trans. JSIDRE*, Vol.124, pp.53-60 (in Japanese).

Takeda, I. & Fukushima, A. (2006). Long-term changes in pollutant load outflows and purification function in a paddy field watershed using a circular irrigation system. *Water Res.*, Vol.40, pp.569–578.

Takeda, I., Fukushima, A. & Tanaka, R. (1997). Non-point pollutant reduction in a paddy-field watershed using a circular irrigation system. *Water Res.*, Vol.31, pp.2685-2692.

Zhou, S. & Hosomi, M. (2008). Nitrogen transformations and balance in a constructed wetland for nutrient–polluted river water treatment using forage rice in Japan. *Ecol. Eng.*, Vol.32, pp.147–155.

Zulu, G., Toyota, M. & Misawa, S. (1996). Characteristics of water reuse and its effects on paddy irrigation system water balance and the riceland ecosystem. *Agric. Water Manage.*, Vol.31, pp.269–283.

7

The Response of Ornamental Plants to Saline Irrigation Water

Carla Cassaniti[1], Daniela Romano[1] and Timothy J. Flowers[2]
[1]Department of Agriculture and Food Science, University of Catania,
[2]School of Plant Biology, Stirling Highway, Crawley, Western Australia,
[1]Italy
[2]Australia

1. Introduction

Salinity affects about one third of irrigated land, causing a significant reduction in crop productivity (Flowers & Yeo, 1995; Ravindran et al., 2007). For this reason researchers have paid considerable attention to this important environmental problem over the last decades. Few studies, however, have dealt specifically with ornamental plants used in landscapes, despite the fact that salt stress causes serious damage in these species (Cassaniti et al., 2009a; Marosz, 2004). Salinity is of rising importance in landscaping because of the increase of green areas in the urban environment where the scarcity of water has led to the reuse of wastewaters for irrigation (McCammon et al., 2009; Navarro et al., 2008). Salinity is also a reality in coastal gardens and landscapes, where plants are damaged by aerosols originating from the sea (Ferrante et al., 2011) and in countries where large amounts of de-icing salts are applied to roadways during the winter months (Townsend & Kwolek, 1987).

Although water is used for purposes other than irrigation, "a landscape may serve as a visual indicator of water use to the general public due to its visual exposure" (Thayer, 1976). While in the past only good quality water (in some States of the USA, homeowners used approximately 60% of potable water to irrigate landscapes; Utah Division of Water Resources, 2003) was used for landscaping and/or floriculture (Tab. 1), nowadays the ecological sensitivity widely diffused in landscape management and planning (Botequilla Leitão & Ahern, 2002) determines the need to explore alternative water sources for irrigation. Landscape water conservation consequently requires making choices of plant species able to tolerate salt stress in order to allow the use of low quality water.

Alternative water sources might be recycled water, treated municipal effluent and brackish groundwater, all of which generally have higher levels of salts compared with potable waters (Niu et al., 2007b). Treated effluent may also contain nutrients essential for plant growth; if water quality is good (not too saline), treated effluent can improve plant growth and reduce fertilizer requirements (Gori et al., 2000; Quist et al., 1999); application of industrial and municipal wastewater to land can be an environmentally safe water management strategy (Rodriguez, 2005; Ruiz et al., 2006). The potential physical, chemical or biological problems that are associated with effluent water applied to edible crops (Kirkam, 1986) are of lesser concern for landscape plant production (Gori et al., 2000).

Characteristics	Desired Level
Soluble salts (EC)	less than 0.5 dS m^{-1}
pH	5.0 to 7.0
Alkalinity (expressed as calcium carbonate)	between 40 and 100 ppm (0.80 and 2.00 me/L^{-1})
Nitrate (NO$_3$)	less than 5 ppm
Ammonium (NH$_4$)	less than 5 ppm
Phosphorous (P)	less than 5 ppm
Potassium (K)	less than 10 ppm
Calcium (Ca)	less than 120 ppm
Sulfates (SO$_4$)	less than 240 ppm
Magnesium (Mg)	less than 24 ppm
Manganese (Mn)	less than 2 ppm
Iron (Fe)	less than 5 ppm
Boron (B)	less than 0.8 ppm
Copper (Cu)	less than 0.2 ppm
Zinc (Zn)	less than 5 ppm
Aluminum (Al)	less than 5 ppm
Molybdenum (Mo)	less than 0.02 ppm
Sodium (Na)	less than 50 ppm
SAR*	less than 4 ppm
Chloride (Cl)	less than 140 ppm
Fluoride (F)	less than 1 ppm

*SAR (Sodium Absorption Ratio) relates sodium to calcium and magnesium levels.

Table 1. Desiderable characteristics of high-quality irrigation water (Source: Dole & Wilkins, 1999).

However, any negative effects of salts on plant growth have to be taken into consideration mainly for their influences on aesthetic value which is an important component of ornamental plants. Salt tolerance does, however, vary considerably among the different genotypes of ornamentals used in landscaping. Ornamental plants can be considered all the species and/or varieties that provide aesthetic pleasure, improve the environment and the quality of our lives (Savé, 2009). This definition is, however, rather imprecise because these plants are used around the world and consequently the concept of 'ornamental' is ambiguous because it includes very important cultural differences (Savé, 2009). Ornamental plants are also used to restore disturbed landscapes, control erosion and reduce energy and water consumption, to improve the aesthetic quality of urban and rural landscapes, recreational areas, interiorscapes and commercial sites. So the number of plant species is very large due to the great geographical range over which they are used and their different functions. In relation to this high number of species that can potentially be utilized in the landscape, the possibility of finding genotypes able to cope with salt stress is high. Unlike in agriculture, performance of an amenity landscape is not measured with a quantifiable yield but how well it meets expectations of the user or the individual paying for installation and maintenance, who may or not be one and the same person. Expectations include aesthetic appearance and/or utility, such as shading, ground cover and recreation (Kjelgren et al., 2000). Sometimes in marginal conditions plant survival is often the only aim of cultivation. Furthermore, for landscape plants, maximum growth is not always essential and indeed excessive shoot vigor is often undesirable. To keep a compact growth habit, ornamentals often have to be pruned or treated with growth regulators (Cameron R.W.F. et al., 2004) so using an alternative water source may be prove advantageous where a more compact form arises as result of salt stress and where slower growth is desiderable for easier landscape management (Niu et al., 2007b). Hence, the use of reclaimed water could conserve potable water and irrigation budgets (Fox et al., 2005). However, to expand the use of such waters while minimizing salt damage, the salt tolerance of ornamentals needs to be determined (Niu & Rodriguez, 2006b).

Apart from plant characteristics, soil composition and drainage characteristics also need to be taken into consideration as they can influence the severity of plant damage by saline irrigation water. For example, clay soils and soils with a high percentage of organic matter exhibit faster and greater build up in concentration of sodium than sandy soils (Dirr, 1976). High concentrations of sodium can displace calcium and magnesium ions, whereas bicarbonate ions can destroy soil structure. This is especially important when irrigation water with high soluble salts is applied on a long-term basis (Fox et al., 2005).

With this in mind the present chapter analyses this large environmental issue as it relates to the response of ornamental plants (herbaceous annuals and perennials, shrubs and woody trees) to salt. We look at the range of tolerance, the possible management practices that could be used to realize a sustainable landscape in which saline water is used and the means available to reduce the effect of salt stress: we also consider the choice of plant species and tailoring plant management to the saline conditions.

2. The response of ornamental plants to salt stress

Salt effects on plants are the combined result of the complex interaction among different morphological, physiological and biochemical processes (Fig. 1).

One of the first responses of plants to salinity is a decreased rate of leaf growth (Blum, 1986) primarily due to the osmotic effect of salt around the roots, which leads to a reduction in water supply to leaf cells. High external salt concentrations can also inhibit root growth (Wild, 1988), with a reduction in length and mass of roots (Shannon & Grieve, 1999) and of function. Reduction in cell elongation and division in leaves reduces their final size, resulting in a decrease in leaf area (Alarcón et al., 1993; Matsuda & Riazi, 1981; Munns & Tester, 2008). Leaf area reduction could be caused by a decrease in turgor in the leaves, as a consequence of changes in cell wall properties or a reduction in photosynthetic rate (Franco et al., 1997). Such consequences are seen in ornamental plants: Cassaniti et al. (2009b) showed that the decrease in shoot dry weight and leaf area were the first visible effects of salinity both in sensitive and tolerant species such as *Cotoneaster lacteus* and *Eugenia myrtifolia*, respectively (Fig. 2). Another common response to high salt level is leaf thickening, which occurred in ornamental plants such as *Coleus blumei* and *Salvia splendens* (Ibrahim et al., 1991).

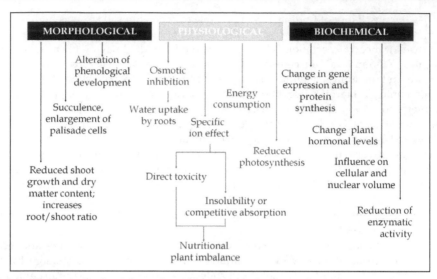

Fig. 1. Morphological, physiological, and biochemical effects of salt stress on plants (modified from Singh & Chatrath, 2001).

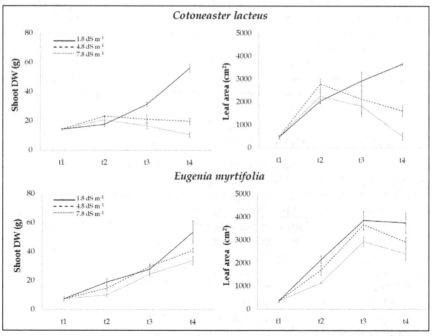

Fig. 2. Shoot dry weight (g) and leaf area (cm²) per plant for *Cotoneaster lacteus* (upper panels) and *Eugenia myrtifolia* (lower panels) at three salinities - 1.8, 4.8 and 7.8 dS m⁻¹ - at the beginning (t₁) and 8 (t₂), 16 (t₃) and 24 (t₄) weeks after the beginning of the salt treatment (Source: Cassaniti et al., 2009b).

Depending upon the composition of the saline solution, ion toxicities or nutritional deficiencies may also reduce growth because of competition between cations or anions (Shannon & Grieve, 1999). When toxic ions such as Na^+ and Cl^- are present in the rhizosphere, they can disrupt the uptake of nutrients by interfering with transporters in the root plasma membrane, such as those for K^+ and NO_3^- (Tester & Davenport, 2003). The influence of salt stress on plant growth alone is, however, not sufficient to evaluate the salt tolerance of ornamentals: tip and marginal leaf burn as consequence of ion toxicity have to be considered (Francois, 1982) due to their influence on decorative value (Fig. 3).

Chloride toxicity manifests as slight bronzing and leaf-tip yellowing followed by tip death and general necrosis, whereas Na^+ toxicity starts as a marginal yellowing followed by a progressive necrosis (Ferguson & Grattan, 2005; Marschner, 1995). Thus the overall appearance as well as survival should be the ultimate criteria governing the choice of landscape species (Townsend, 1980). Many methodologies based on visual quality ratings have been developed by different authors to evaluate the appearance of ornamental plants in response to salt stress (Tab. 2).

Fig. 3. Necrotic areas due to the effect of salt stress on some ornamental species: a) *Cotoneaster lacteus*; b) *Grevillea juniperina* var. *sulphurea*; c) *Pyracantha* 'Harlequin'; d) *Teucrium fruticans* (Source: Cassaniti, 2008).

Source	Rating/Marks	Considered attributes
Cassaniti, 2008	1= no leaf necrotic area; 2= leaf necrotic area between 0 and 33%; 3= leaf necrotic area between 33 and 66%; 4= necrotic area between 66 and 100%	Incidence of leaf necrosis: percentage of necrotic areas, leaf bronzing
Fox et al., 2005	1= dead plant; 2= severe damage such as stunting, dead stems; 3= moderate damage such as visible salt residue on the foliage, < 50% defoliation, leaf deformity, necrosis; 4= slight damage, such chlorosis, tip and/or marginal leaf burn, spotting; 5=no damage, highest aesthetic quality	Stunting, discoloration, defoliation
Jordan et al., 2001	Each parameter was evaluated on a 1-9 scale, where a value of 1 equated to a rating of 10% and a value of 9 equated to a rating of 90% damage	Absence of crown dieback, overall canopy discoloration, presence of dead leaves, presence of deformed leaves, discoloured leaves and tip and marginal damage
Niu & Rodriguez, 2006a, 2006b	0= dead; 1= severely stunted growth with over 50% foliage salt damage (leaf necrosis, browning); 2= stunted growth with moderate (25-50%) foliage salt damage; 3= average quality with slight (<25%) foliage salt damage; 4= good quality with acceptable growth reduction and little foliage damage; 5= excellent with vigorous growth with no foliage damage	Leaf necrosis, browning
Niu et al., 2007a, 2007b	1= over 50% foliage damage or plant dead; 2= moderate foliage damage (25-50%); 3= slight foliage damage <25%); 4= good quality with acceptable growth reduction and little foliage damage; 5= excellent with no foliage damage	Salt damage: burning and discoloration
Quist et al., 1999	Ranking scale of 1 to 10 with a score of 10 indicating plants with the highest quality	Necrosis, chlorosis, leaf color, leaf turgor, tip die back, misformed leaves, leaf size, leaf loss and disease.
Valdez-Aguilar et al., 2011	1= poor quality, leaf bronzing higher than 75% or dead plants; 5= best quality	Leaf bronzing, leaf scorching, overall appearance
Zollinger et al., 2007	Salt damage: 1= more than 50% of leaf area damaged ; 2= 25% to 50% of the leaf area damaged; 3= 5% to 24% of the leaf area damaged; 4= less than 5% of the leaf area damaged with burn or discoloration primarily restricted to leaf damage. Wilt: 1= more than 65% of the plant was wilted; 2= 35% to 65% of the plant was wilted; 3= 5% to 34% of the plant was wilted; 4= less than 5% of the plant was wilted	Salt damage: burning/ discoloration, wilting

Table 2. Some visual quality rating scales for evaluating salt damage on the foliage.

The tolerance to saline water can also be evaluated by growth analysis indices: for example, plant response to water deficit and saline treatment was investigated by Rodríguez et al. (2005) using *Asteriscus maritimus*, a native species of coastal areas. Salinity caused a reduction of RGR (Relative Growth Rate) and NAR (Net Assimilation Rate) at 70 mM and 140 mM NaCl (about 7 and 14 dS m^{-1}, respectively) while LAR (Leaf Area Ratio) was not affected. However, the LWR (Leaf Weight Ratio) increased in plants treated at 140 mM NaCl, due to the greater reduction of stem than leaf dry weight. LWR is an important parameter for ornamental plants, in which the aesthetic value is strictly correlated to the appearance of the leaves. RGR reduction clearly suggests a direct effect of the stress on stomatal closure and/or photosynthetic apparatus, indicating that photosynthesis could be the growth-limiting factor (Cramer et al., 1990; Sánchez-Blanco et al., 2002).

Herbaceous, annuals and perennials, show different responses to salinity than woody plants, although similar mechanisms can be involved. Because a typical landscape is a blend of species (annuals, grasses, climbing plants, shrubs, trees and palms; Graf, 1992) it is important to determine the salt tolerance of all commonly used plants in any specific landscape to minimize potential salt damage before converting to treated effluents or any other non-potable water source (Niu & Rodriguez, 2006a). Evaluating salt tolerance is made more complex by intra-specific variation: any given species can vary in its tolerance to salinity, depending on genotype.

2.1 Herbaceous plants

Herbaceous perennials are popular for landscaping because of their low maintenance and as their planting increases diversity in the landscape (Cameron A. et al., 2000; Johnson & Whitwell, 1997). Herbaceous plants do, however, show a very variable response to salt stress, from the tolerant halophytes to the sensitive glycophytes and their sensitivity to salty irrigation water can influence plant selection, irrigation method and frequency of watering.

The irrigation of landscapes with treated effluent has become a common practice in states like Florida and California (Cuthbert & Hajnosz, 1999; Parnell, 1988) where the municipal water consumption typically increases by 40-60% for landscape irrigation during summer months (Kjelgren et al., 2000) and could be important for Mediterranean countries. Studies on herbaceous perennials in semiarid parts of United States (Niu & Rodriguez, 2006a, 2006b) have involved herbaceous perennials and have emphasized the importance of visual quality for expressing relative salt tolerance and acceptability of species for landscape use (Fox et al., 2005; Niu & Rodriguez, 2006a, 2006b). Research programs conducted in Israel have revealed ornamental species suitable for saline environments or to be irrigated with salt waters (Forti, 1986). Perennial turf grasses have been selected at Arizona University to cope with a salt concentration more than 15 g L^{-1}: *Distichlis spicata*, commonly known as *desert saltgrass*, is able to survive at concentration up to 400 mM of NaCl (Pessarakli et al., 2001).

Niu & Rodriguez (2006b) observed the response to salt stress on eight herbaceous perennials (*Penstemon eatonii, P. pseudospectabilis, P. strictus, Ceratostigma plumbaginoides, Delosperma cooperi, Lavandula angustifolia, Teucrium chamaedrys, Gazania rigens*); three salt treatments plus control were tested (3.2, 6.4, 12 and 0.8 dS m^{-1}). The relative water content significantly declined as salinity increased in *C. plumbaginoides* and in *G. rigens* at the highest salt level; *D. cooperi* showed the highest water potential due to increased succulence of the leaves, a common mechanism of salt tolerance (Kozlowski, 1997). The higher Na$^+$ concentration in roots than shoots indicated in this plant Na$^+$ exclusion from aerial parts and a capability to tolerate Cl- in leaf tissues. Plants of *L. angustifolia* and most of the species of *Penstemon* showed symptoms of necrosis and eventually died with an EC more than 3.2 dS m-1. Among the *Penstemon* species, the earliest leaf injury appeared on *P. strictus*, perhaps related to its rosette growth habit allowing some leaves to have been in direct contact with saline water during irrigation. *G. rigens* did not show any injury symptom even at the highest salt level, although growth was stunted growth, probably due to the high Na+ accumulation in the shoots. *T. chamaedrys* exhibited necrosis in some leaves at medium and high salt level, while *C. plumbaginoides* manifested slight leaf browning at 3.2 dS m-1, severe symptoms at 6.4 dS m-1 and death of many plants at 12 dS m-1.

As we have noted, visual quality is an important factor in the choice of herbaceous perennials for saline landscapes (Tab. 2). Because plants respond differently to salinity, visual quality may or may not be related to biomass production and photosynthetic response (Zollinger et al., 2007). Following this argument, Niu & Rodriguez (2006b) argued that *Gazania rigens* and *Delosperma cooperi* can be used in a landscape irrigated with saline waters as, in spite of their decrease in growth rate, they did not show any injury symptoms. The other species they tested could be considered salt sensitive, since most of them died at the highest salt treatments. *L. angustifolia* began to show leaf injury about 4 weeks after the start of saline irrigation, and died in the subsequent weeks at 6.4 and 12 dS m-1. However, the importance of the nature of the salts present is illustrated by the results of Zollinger et al. (2005), who reported that *L. angustifolia* survived at 8.3 dS m-1 when NaCl and CaCl$_2$ (2:1 molar ratio) were used for saline solution.

Earlier research indicated that the climatic conditions can also influence the extent of foliar damage (Jordan et al., 2001; Quist et al., 1999; Wu et al., 1999). Subsequently various trials have been reported comparing data obtained by conducting experiments in different seasons and in years when the climatic conditions varied considerably. For example, a trial was conducted for testing the salinity tolerance of five herbaceous perennials commonly used in the landscape, *Achillea millefolium, Agastache cana, Echinacea purpurea, Gaillardia aristata* and *Salvia coccinea* (Niu & Rodriguez, 2006a). In this case, the tolerance to salt stress was evaluated, in summer and fall, with many parameters - dry weight, plant height, osmotic potential and visual score – being used to estimate the damage to their ornamental value (Tab. 2). Plants were treated with three salinity levels: 0.8, 2 and 4 dS m-1. In the summer experiment, all species showed a lower osmotic potential at 2 dS m-1 and 4 dS m-1 compared the control. Neverthless, despite the reduction in dry weight that occurred in salt treatments, *A. millefolium, G. aristata* and *S. coccinea* showed a visual score acceptable for landscape performance. When the experiment was conducted in the fall, the lowering of the osmotic potential in these species was much less than occurred during the summer. As confirmed by other authors (Niu et al., 2007a; Zollinger et al., 2005), results highlighted how environmental conditions could influence the response to salt stress: the higher temperature and irradiance typical of summer meant that plants became more stressed than in the fall, when all the species, except *A. cana*, maintained an acceptable visual quality. Species like *S. coccinea, A. millefolium* and *G. aristata* were considered highly salt tolerant, because they could be irrigated with a saline solution up to 4 dS m-1 under both summer and fall conditions, with little or no growth reduction.

In experiments again conducted during different seasons (spring, summer or fall) at different salinities (0.3, 1.9, 5.0 and 8.1 dS m-1) were tested on eight species (Zollinger et al.,

2007). Species were selected for being native of inter-mountain Western United States (*Penstemon palmeri, Mirabilis multiflora, Geranium viscosissimum, Eriogonum jamesii*) or available in the nursery industry (*Echinacea purpurea, Lavandula angustifolia, Leucanthemum ×superbum* 'Alaska' and *Penstemon ×mexicali* 'Red Rocks'). Light intensities and greenhouse temperatures varied among the seasons and had an impact on the response of certain species to salinity. Results suggested that irrigation with saline water would lower the visual quality of *G. viscosissum, E. purpurea* and *P. palmeri,* more during the warmer, summer months than at cooler times of the year.

Based on visual score, Fox et al. (2005) evaluated the response to treated effluent as an irrigation source (from 0.75 dS m^{-1} to 2.5 dS m^{-1}) of seven annuals and seven perennials in a two-year experiment when conditions differed considerably. Damage symptoms were more severe in 2001 than 2000, which was characterized by having the hotter and drier summer, confirming the important influence of temperature on plant performance under saline conditions.

Another important aspect of salinity in the landscape is the foliar absorption of ions, whether from irrigation water or aerosols produced by wind blowing over seawater. Plant species typical of the coastal areas have adapted to survive direct contact of the salt on the leaves, although the exposure to sea aerosol and salt water infiltration of the ground water may well reduce plant growth and affect their reproduction (Cheplick & Demetri, 1999; Hesp, 1991). The presence of surfactant can, however, enhance the foliar absorption of sea salt through stomatal and cuticular penetration (Greene & Bukovac, 1974; Schönherr & Bauer, 1992). Sánchez-Blanco et al. (2003) conducted a trial to evaluate the response to sea aerosol of two wild native species from littoral areas, *Argyranthemum coronopifolium* and *Limonium pectinatum*. Plants were treated with one of three solutions: one containing an anionic surfactant, one simulating the composition of sea aerosol and a third with sea aerosol and anionic surfactant; the control involved spraying with deionized water alone. The most sensitive to sea aerosol was *A. coronopifolium*, in which salt sprays reduced its growth and dry mass, while any effect on *L. pectinatum* was not evident. Although it is a native plant of coastal areas, *A. coronopifolium* is not a salt-tolerant species (Morales et al., 1998). On the other hand, the halophyte *L. pectinatum* was more tolerant, directly excreting salts from its leaves (Alarcón et al., 1999). Foliar damage is directly linked to foliar absorption, with increased leaf ion penetration with increasing temperature (Darlington & Cirulis, 1963).

2.2 Shrubs and woody trees

Although there have been many studies on the effects of salt stress on landscape plants, few have investigated the effects of salinity on shrubs, despite their importance for landscaping (Bernstein et al., 1972; Bañon et al., 2005; Cassaniti et al., 2009a; Francois & Clark, 1978; Picchioni & Graham, 2001; Valdez-Aguilar et al., 2011) and production in Mediterranean countries (Marosz, 2004; Zurayk et al., 1993). Salinity may affect the growth of ornamental shrubs by reducing growth and leaf expansion resulting from osmotic effects or toxicity due to the high concentration of Na$^+$ and Cl$^-$ typical of saline water (USEPA, 1992). In a study conducted on 15 ornamental shrubs commonly used for landscaping (Cassaniti, 2008), many parameters (dry weight of different organs, leaf area, number of leaves, SPAD, growth indexes, aesthetic value) were considered to evaluate plant response to salt stress. Plants were grown in a greenhouse and subjected for a six month period to three salt levels: 1.8, 4.8 and 7.8 dS m⁻¹. The aesthetic value, calculated as percentage of leaves showing necrotic areas (Tab. 2), was affected by the increasing EC, confirmed by the increased percentage of leaf necrosis, most of all in *Cotoneaster lacteus* and *Grevillea juniperina* var. *sulphurea*. These symptoms were clearly associated with the large amount of Na^+ and Cl^- in the leaves (Cassaniti et al., 2009a; Karakas et al., 2000). Shoot dry weight was reduced in many species and, in general terms, these reductions followed the same trend as the leaf area. Leaf number was affected by salinity, hence leaf abscission reduced the photosynthetic area (c.f. Munns & Termatt, 1986). Chlorophyll content (SPAD index) and root dry weight were less influenced than the other parameters by salt treatment at the end of the experimental period. Among the many parameters analysed, the relative growth rate and shoot dry weight best highlighted the differential response to salt stress (Fig. 4). Therefore, based on the shoot dry weight reduction, plants were grouped in four categories: (1) salt sensitive species, showing more than 75% reduction (*Cotoneaster lacteus*, *Pyracantha* 'Harlequin'); (2) moderately salt sensitive species, showing between 50 and 75% growth reduction (*Grevillea juniperina* var. *sulphurea*); (3) moderately salt tolerant species, showing a growth reduction between 25 and 50% (*Cestrum aurantiacum* and *Cestrum fasciculatum* 'Newellii' *Escallonia rubra*, *Viburnum lucidum*, *Teucrium fruticans*, *Eugenia myrtifolia*, *Ceanothus thyrisiflorus* var. *repens*, *Bougainvillea glabra*, *Ruttya fruticosa*, *Polygala myrtifolia*); (4) salt tolerant species, showing less than 25% growth reduction (*Leucophyllum frutescens*, *Leptospermum scoparium*).

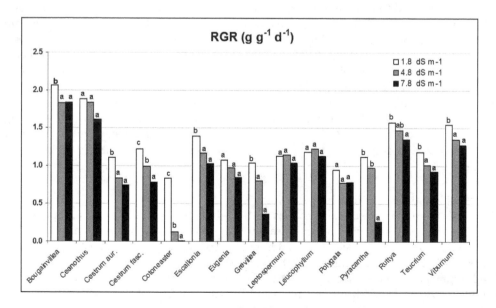

Fig. 4. RGR (Relative Growth Rate g g⁻¹ d⁻¹) of ornamentals shrubs in relation to three salt levels (1.8, 4.8 and 7.8 dS m⁻¹) calculated between the beginning and end of experimental period (180 days) (Source: Cassaniti, 2008).

As observed for herbaceous species, the growing season seems to affect the response of shrubs to salt (Valdez-Aguilar et al., 2011). Five species (*Buxus microphylla* var. *japonica*, *Escallonia* ×*exoniensis* 'Fradesii', *Raphiolepis indica* 'Montic', *Hibiscus rosa-sinensis* 'Brilliant' and *Juniperus chinensis* 'Torulosa') were investigated at different salinities (EC of irrigation water of 0.6, 2, 4, 6 and 8 dS m^{-1}) for two growing seasons (starting from May and September, respectively, until plant harvest). Species were ranked for their salinity tolerance according to the slope of linear regressions of reduction in leaf DW as the EC increased; tolerances were, from higher to lower: *H. rosa-sinensis*, *J. chinensis*, *E.* ×*exoniensis*, *R. indica* and *B. microphylla*. Plant response differed with growing season. Roots of *R. indica*, *J. chinensis* and *E.* ×*exoniensis* accumulated more DW during the spring-summer while leaves accumulated more DW in the fall-winter. The highest root DW recorded, during the spring-summer cultivation, allowed the allocation of Ca^{2+}, Na$^+$ and Cl$^-$ to the roots, preventing toxic concentrations accumulating in the leaves. This mechanism, Na$^+$ and Cl$^-$ retention by the roots, was especially efficient in *Juniperus*, which was rated as one of the most tolerant in terms of growth and visual quality. However, as we have mentioned previously, for ornamental plants responses other than DW need to be considered, because plants with good DW accumulation can still show a high percentage of leaf bronzing, while other species respond with high reductions in shoot dry weight but no visual damage. *B. microphylla* exhibited acceptable tolerance in terms of growth but the visual quality of the final product was impaired. Growth of *Hibiscus* was the most severely reduced when grown in spring-summer but the lack of injury in leaves suggests salt compartmentalization (Rodríguez et al., 2005; Sánchez-Blanco et al., 2004).

Apart from reductions in DW, other reported effects of salinity on woody species that affect their visual appearance include crown dieback, lesions on the stem or trunk and leaf scorch (Percival, 2005). Buds may fail to open or grow and branches may die. The morphological adaptations allowing trees to cope with salinity can include penetration-resistant resinous buds and waxy leaves and stems. Mechanisms of salt exclusion can be smooth twigs, sunken buds and low surface-to-volume ratios (e.g., pine needles) (Appleton et al., 1999). On conifers, damage appears as brown needle tips (Azza Mazher et al., 2007). With time, symptoms may accumulate causing tip burn of the older needles in conifers with their consequent necrosis, die back of limbs and death (Dobson, 1991). Salt damage on evergreen trees usually first appears in late winter to early spring and becomes more extensive during the growing season (Azza Mazher et al., 2007). A general crown thinning may also occur as salt build-up in the soil causes soil structure to deteriorate and roots to be damaged. Trees may become misshapen due to greater damage on the side facing the wind or where trees stand taller than partially protective buildings (Appleton et al., 1999). NaCl applied to the canopy of *Thuja occidentalis* and *Picea glauca* induced fragmented cuticles, disrupted stomata, collapsed cell walls, coarsely granulated cytoplasm, disintegrated chloroplasts and nuclei and disorganized phloem (Kozlowski, 1997). A reduction in branch diameter may also occur (Stroganov, 1964).

As for herbaceous plants, the ionic composition of the irrigation water can affect the response of shrubs and trees to saline stress. Chloride salts seem to be more damaging than SO$_4^{2-}$ salts, and Mg^{2+} associated with Cl$^-$ is more damaging than Na$^+$ with Cl$^-$ (Devitt et al., 2005a). However, trees like *Eucalyptus occidentalis* and *E. sargentii* can tolerate salinity of about 30 dS m^{-1} (Choukr-Allah, 1997). Some results on salt response of ornamental shrubs are listed in the table below (Tab. 3).

Species	Rating	Salt response	Salinity threshold	References
Bougainvillea spectabilis, Lantana camara var. *aculeata*	Tolerant	Maintains a high visual quality	1.94 dS m^{-1}	Devitt et al., 2005b
Poinciana pulcherrima	Questionable	Little foliar damage	1.94 dS m^{-1}	Devitt et al., 2005b
Euonymus japonica, Fraxinus pennsylvanica var. *lanceolata, Taxus cuspidata, Tilia europaea*	Sensitive	Low rank score in visual quality	2.1 dS m^{-1}	Quist et al., 1999
Gleditsia tricanthos var. *inermis, Prunus cerasifera* var. *atropurpurea, Berberis thunbergii* var. *atropurpurea, Pinus nigra, Pyrus calleryana, Picea pungens, Juniperus chinensis* var. *pfitzeriana*	Tolerant	High rank score in visual quality	2.1 dS m^{-1}	Quist et al., 1999
Crataegus opaca	Sensitive	Reduction in relative growth rate (RGR)	3.15 dS m^{-1}	Picchioni & Graham, 2001
Olea europaea 'Swan Hill', *Prosopsis chilensis, Pinus halepensis, Pinus eldarica, Rhus lancea, Pinus pinea, Fraxinus oxycarpa* 'Raywood'	Tolerant	Good visual quality	1.87 dS m^{-1}	Jordan et al., 2001
Robinia ×*ambigua* 'Idahoensis', *Vitex agnus-castus, Quercus virginiana* 'Heritage', *Albizia julibrissin*	Questionable	Medium visual quality	1.87 dS m^{-1}	Jordan et al., 2001
Salix matsudana 'Navajo', *Prunus cerasifera* 'Atropurpurea', *Cercidium floridum, Ligustrum japonicum, Chitalpa tashkentensis* 'Pink Dawn', *Ulmus parvifolia* 'Drake', *Chilopsis linearis, Pistacia chinensis, Fraxinus velutina* var. *glabra* 'Modesto'	Sensitive	Low visual quality	1.87 dS m^{-1}	Jordan et al., 2001
Lantana ×*hybrida* 'New Gold', *Lonicera japonica* 'Halliana', *Rosmarinus officinalis* 'Huntington Carpet'	Tolerant	Little reduction in growth, good aesthetic appearance	5.4 dS m^{-1}	Niu et al., 2007a
Lantana montevidensis	Sensitive	Reduction in growth index, low aesthetic appearance	5.4 dS m^{-1}	Niu et al., 2007a
Potentilla fruticosa 'Longacre', *Cotoneaster horizontalis*	Tolerant	No growth reduction and visible effects	12 dS m^{-1}	Marosz, 2004
Cotoneaster 'Ursynów', *Spiraea* 'Grefsheim'	Sensitive	Leaf injuries	12 dS m^{-1}	Marosz, 2004
Arbutus unedo	Sensitive	Reduction of total dry biomass	5.45 dS m^{-1}	Navarro et al., 2007

Table 3. Results of studies of ornamental shrubs in order to evaluate salt response.

3. Salt tolerance in ornamentals

3.1 Assessment of tolerance

Plant salt tolerance is the ability to withstand the effects of high salt concentrations in the root zone without a significant adverse effect (Shannon & Grieve, 1999). Maas and Grattan (1999) grouped crop species into five or six salt tolerance divisions based on growth, but for plants used in landscaping this is not necessarily the best approach; a separation based on the visual quality of the plants is often the most useful. A further complication to the assessment of tolerance is the fact that plants are generally more severely injured by saline water applied by sprinkler than by drip systems (Maas & Francois, 1982). Saline water applied by sprinklers can coat plant foliage burning and desiccating the leaves of sensitive species (Fox et al., 2005), although sometimes a waxy cuticle on the leaves can make them less sensitive to aerial salt than to soil salt (e.g. on plants native to strand lines and of some woody species; Van Arsdel, 1996). Furthermore, where grafted trees are used in landscaping – generally fruit trees - the genetic differences between rootstock and scion can confound evaluation of relative tolerance to sprinklers and drippers (Musacchi et al., 2006).

Despite these complexities, field trials have been conducted to compare sprinkler and drip irrigation systems able to differentiate salt resistance among landscape species based on their aesthetic quality (Miyamoto et al., 2004; Wu et al., 2001a, 2001b). For example, a study conducted in California (Wu et al., 2001b) on ten ornamentals used in the landscape (*Pistacia chinensis, Nerium oleander, Pinus cembroides, Buxus microphylla, Liquidambar styraciflua, Bignonia violacea, Ceanothus thyrsiflorus, Nandina domestica, Rosa* sp., *Jasminum polyantum*) confirmed that the species showed a higher sensitivity when irrigated with sprinkler than drip irrigation. In studies conducted on native species of coastal areas other authors observed a large variation in foliage damage (with no symptoms to severe injury) in species that showed similar salinity tolerance in the roots (Cartica & Quinn, 1980; Sykes & Wilson, 1988), confirming that plants can evolve resistance to saline aerosols.

In a further trial, Wu et al. (2001b) evaluated 38 trees and ten herbaceous perennial subjected to two salt levels: 500 mg L^{-1} NaCl (200 mg L^{-1} Na$^+$, 300 mg L^{-1} Cl$^-$) and 1500 mg L^{-1} NaCl (600 mg L^{-1} Na$^+$, 900 mg L^{-1} Cl$^-$) to evaluate those that could be irrigated with brackish water (Tab. 4).

Results showed that 21 (55%) of the 38 woody plant species and 7 (70%) of the 10 native grass species were salt tolerant when irrigated with 500 mg L^{-1} salt and twelve (31%) woody species and 5 (50%) grass species were salt tolerant when they were irrigated with 1500 mg L^{-1} salt. Wu & Dodge (2005) summarized the results of previous trials, listing the salt tolerance of 268 species (72 trees, 15 palms, 66 shrubs, 39 ground covers, 18 vines, 58 grasses) used in landscaping, based on plant response to salt applied with sprinklers or through soil. Tolerances to salt spray were defined by the degree of visual damage on the leaves (relative to plants irrigated with potable water) and the salt concentrations in the applied irrigation water, while tolerances to soil salinity were defined as the limit of soil salinity that did not induce significant salt stress symptoms. Species were grouped in four categories: highly tolerant, tolerant, moderately tolerant, and sensitive (Tab. 5). Generally the species that tolerate salt spray tolerate soil salinity. Approximately 50% of landscape ornamentals are either tolerant or moderately tolerant to salt.

Scientific name	Tolerance to NaCl		Scientific name	Tolerance to NaCl	
	500 mg L^{-1}	1500 mg L^{-1}		500 mg L^{-1}	1500 mg L^{-1}
Woody landscape plants			*Nerium oleander*	High	High
Abelia ×grandiflora 'Edward Goucher'	Low	Low	*Olea europaea* 'Montra'	High	High
Acacia redolens	High	High	*Pinus cembroides*	High	High
Albizia julibrissin	Moderate	Low	*Pistacia chinensis*	Low	Low
Arbutus unedo	High	Moderate	*Pittosporum tobira*	High	High
Buddleja davidii	Low	Low	*Plumbago auriculata*	High	High
Buxus japonica	High	High	*Prunus caroliniana*	High	Low
Ceanothus thyrsiflorus	High	Moderate	*Quercus agrifolia*	High	Moderate
Cedrus deodara	High	High	*Rhaphiolepis indica*	High	High
Celtis sinensis	Low	Low	*Rosa* sp.	Low	Low
Clytostoma callistegioides	Low	Low	*Sambucus nigra*	Moderate	Low
Cornus mas	Low	Low	*Sapium sebiferum*	High	High
Cotoneaster microphyllus 'Rockspray'	Moderate	Low	*Washingtonia filifera*	High	High
Escallonia rubra	High	Moderate			
Euryops pectinatus	Low	Low	Herbaceous landscape plants		
Forsythia ×intermedia	High	Moderate	*Bromus carinatus*	High	Moderate
Fraxinus angustifolia	Moderate	Low	*Deschampsia cespitosa*	Moderate	Low
Ginkgo biloba	Low	Low	*Deschampsia elongata*	High	Moderate
Jasminum polyanthum	High	Moderate	*Elymus glaucus*	High	High
Juniperus virginiana 'Skyrocket'	High	High	*Festuca californica*	High	High
Koelreuteria paniculata	Moderate	Low	*Melica californica*	Low	Low
Lantana camara	High	Moderate	*Muhlenbergia rigens*	High	High
Liquidambar styraciflua	Low	Low	*Poa scabrella*	Moderate	Low
Mahonia pinnata	Moderate	Low	*Sporobolus airoides*	High	High
Myrtus communis	High	Moderate	*Stipa pulchra*	High	High
Nandina domestica	Moderate	Low			

Table 4. List of salt tolerance of 38 landscape woody plant species and ten California native grass species grown under sprinkler irrigation with two NaCl concentrations (Source: Wu et al., 2001b).

Degree of tolerance	Salinity	
	Spray	Soil
Highly tolerant (H)	No apparent salt stress symptoms were observed when the plants were irrigated with water having 600 mg L^{-1} sodium and 900 mg L^{-1} chloride (salt concentrations rarely reach these levels in recycled water)	Acceptable soil electrical conductivity (EC) greater than 6 dS m^{-1} and plants may not develop any salt stress symptoms even if the soil salinity exceeds this permissible level
Tolerant (T)	No apparent salt stress symptoms were observed when the plants were irrigated with water having 200 mg L^{-1} sodium and 400 mg L^{-1} chloride	Acceptable EC greater then 4 and less than 6 dS m^{-1} and the plants in this category are adaptable to most reclaimed water irrigation without extra management input if restricted to soil application
Moderately tolerant (M)	Salt stress symptoms were observed in 10% or less of leaves when the plants were irrigated with water having 200 mg L^{-1} sodium and 400 mg L^{-1} chloride under dry and warm weather conditions	Acceptable EC greater than 2 and less than 4 dS m^{-1}, plants in this category require extra irrigation and soil management input
Sensitive (S)	Salt stress symptoms were seen in 20% or more of leaves when the plants were irrigated with water having 200 mg L^{-1} sodium and 400 mg L^{-1} chloride.	Acceptable EC less than 2 dS m^{-1} and plants in this category are very sensitive to soil salinity

Table 5. Definitions of salt tolerance categories for the plant species subjected to salt spray and soil salinity (Source: Wu & Dodge, 2005).

3.2 Mechanisms of tolerance

To cope with salinity, plants trigger divergent mechanisms that allow their adaptation and survival in saline environments; differences in the mechanisms determine their performance under saline conditions (Paranychianakis & Chartzoulakis, 2005).

Among the many mechanisms of salinity tolerance (Munns & Tester, 2008), the ability to restrict the entry of saline ions through the roots and limit the transport of Na$^+$ and/or Cl$^-$ to aerial parts, retaining these ions in the root and lower stem, has to be one of the most important of all the traits associated with tolerance (Colmer et al., 2005; Maathuis & Amtmann, 1999; Murillo-Amador et al., 2006). A related trait, the retention of toxic ion in roots, has been proposed to be important to salt tolerance in plants (Boursier & Läuchli, 1990; Pérez-Alfocea et al., 2000). Species that keep acceptable growth rates and possess mechanisms to exclude Na$^+$ and Cl$^-$ from roots or leaves and still have good appearance are ideal for landscaping.

As for all species, ornamentals differ in this trait. For example, *Rudbeckia hirta* 'Becky Orange' and *Phlox paniculata* 'John Fanick' accumulated large quantities of Cl$^-$ in the leaves which led to dry weight reduction of about 25%, while *Lantana* ×*hybrida* 'New Gold' and *Cuphea hyssopifolia* 'Allyson' tolerated salinity extremely well showing the low Cl$^-$ accumulation (Cabrera et al., 2006). The low reduction and absence of salt injury symptoms in *Eugenia myrtifolia* has been associated not only with the root storage of Na$^+$ and Cl$^-$ but also with their restricted uptake as the salinity increased (Cassaniti et al., 2009a).

An important aspect of salt tolerance is related to the ability of a plant to compartmentalize toxic ions, such as Na$^+$ and Cl$^-$ (Boursier & Läuchli, 1989). In this sense, species such as

Bougainvillea glabra, Ceanothus thyrsiflorus and *Leucophyllum frutescens* (Cassaniti et al., 2009a) and *Cistus monspeliensis* (Sánchez-Blanco et al., 2004) accumulated high concentrations of Na+ and Cl- in the leaves but without showing any symptoms of necrosis. For woody perennials Cl- is more problematic than Na+ which usually tends to be sequestered in roots and woody tissue (Ferguson & Grattan, 2005; Storey & Walker, 1999), while Na+ seems the primary cause of ion specific damage in grasses (Tester & Davenport, 2003).

Among the factors used to characterize salt tolerance in crop plants, the maintenance of a high K+/Na+ ratio in their tissues is an important diagnostic character (Maathuis & Amtmann, 1999; Munns & James, 2003), due to competitive effect of Na+ concentration in the rhizosphere on K+ uptake (Aktas et al., 2006; Carvajal et al., 1999). However, amongst ornamentals this parameter has rarely, as far as we are aware, been recorded, although a reduction of K+ concentration was detected in leaves of *Arbutus unedo* with an increase of salinity (Navarro et al., 2008).

The adaptability to salt stress can be also different between and within species belonging to the same genus. For example, Sánchez-Blanco et al. (2004) showed that *Cistus monspeliensis* seemed to be more tolerant to aerosol treatment than *C. albidus,* showing a minor reduction in growth and foliar damage. Results were confirmed by Torrecillas et al. (2003) in which the two species, irrigated with saline water, showed different tolerance mechanisms involving Na+ and Cl- inclusion, leaf area reduction and osmotic adjustment. *C. monspeliensis* had a higher water use efficiency than *C. albidus.* Although, some investigations have been carried out in order to assess the difference between species in the same genera of ornamental plants such as *Cestrum* spp. (Cassaniti, 2008), *Cotoneaster* spp. (Marosz, 2004) and *Lantana* spp. (Niu et al., 2007a), no information within cultivars is available.

4. Sustainable landscape

A "sustainable landscape" commonly refers to one that supports environmental quality and conservation of natural resources (Rodie & Streich, 2009). As reported in the Brundtland Report (1987), the concept of sustainability, or the needs of the present without compromising the ability of future generations to utilise the land, is of increasing interest. Other terms such as xeriscape, native landscape, xerogarden, wild garden and environmental friendly landscape have often been used to describe such landscapes (Franco et al., 2006; Rodie & Streich, 2009). A well-designed sustainable landscape reflects a high level of self-sufficiency, even though this can be difficult to achieve due to the environmental stresses and artificial conditions of urban areas. Sustainable landscapes try to: 1) enhance landscape microclimate; 2) increase biodiversity; 3) reduce resource inputs and waste and 4) maximize re-use of resources. The benefits achievable in sustainable landscapes include enhanced beauty, low environmental decline and water consumption, reduction of use of pesticides and of other chemical resources, the generation of valuable wildlife habitats, and cost savings from reduced maintenance, labour and resource use. At the current time, the use of saline water could also be included as a benefit in accordance with the new trend of planning landscape with agronomical, political, social, cultural and ecological needs (Hitchmough, 2004). To realize a sustainable landscape as far as the use of saline water is concerned, two aspects have to be considered: the choice of plant species and tailoring the plant management to reduce the effects of salt stress.

4.1 Choice of species resistant to salt stress

The response to saline water varies greatly among the plants that are used or potentially used in landscape design (see above; Niu et al., 2007b), often an intricate blend of woody and herbaceous ornamentals with a vast array of manufactured elements (generally referred to as 'hardscape'; Iles, 2003). The plant choice can be based on a very large number of species from a wide geographical range and with different functions in the landscape (Savé, 2009) and whose adaptability changes within genera or species (Sànchez-Blanco et al., 2002; Torrecillas et al., 2003). Where salinity is an issue, although many ornamentals are adversely affected, the choice can include plants that grow naturally on coastal and inland saline areas such as salt marshes and salt deserts (halophytes), and survive at salt concentration equal to or greater than that of seawater (Flowers & Colmer, 2008; Ravindran et al., 2007). As well as halophytes, a wide number of ornamental plants are able to tolerate salt stress (see above) so that making the appropriate choices is important to simplify the work (Fig. 5).

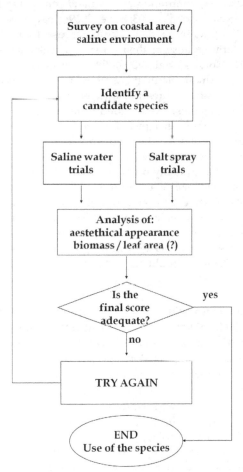

Fig. 5. A schematic summary of the steps included in the selection of a suite of landscape species.

The first step is an exploration of plants growing within seaside and other saline environments; subsequently species have to be tested experimentally to verify at which salt level they are able to cope with salinity without compromising their appearance. In relation to the large number of genotypes potentially available and the difficulty of making choices appropriate to the particular environmental conditions, unsuitable genotypes have to be excluded. Cassaniti & Romano (2011) carried out a survey to identify halophytes, which naturally grow in the Mediterranean area and which could be utilized for ornamental purpose. The investigation, based on the literature, showed about 172 suitable species in 30 families and 86 botanical genera were available. Most suitable species (34) came from the Chenopodiaceae and within the genera, the most represented were *Limonium* (Plumbaginaceae; 17 species) and *Atriplex* (Chenopodiaceae; 14 species).

Native plants can assume an important role for their adaptability to both biotic and abiotic stress (Iles, 2003; Kotzen, 2004; Savè, 2009). Although interest in them has recently risen for restoring disturbed landscapes, controlling erosion and improving the aesthetic quality of environments (Martinéz-Sanchéz et al., 2003; Savè, 2009), they have been largely ignored in landscaping (Romano, 2004). Many of them could represent an alternative to the species traditional used, particularly because of their enhanced water use efficiency (Clary et al., 2004; Franco et al., 2002; Morales et al., 2000). The key question here is can only native species be used sustainably? Undoubtedly there are many alien species that are salt tolerant, but there is a critical necessity to preserve and enhance the ecological and landscape integrity of particular environments, like the Mediterranean or desert areas (Kotzen, 2004). Thus a new landscape paradigm that includes the wide use of native plants has been developed, which appears appropriate for arid and coastal ecosystems where salinity is frequent (Kotzen, 2004; Sánchez-Blanco et al., 2003). However, the introduction of alien species has its dangers because a) many of the introduced species will not survive without large inputs of water and maintenance and b) those that are xerophytes or halophytes and can adapt to the drought/salt condition can turn out to be pernicious weeds (Kotzen, 2004). Hence the threat from using exotic plants needs to be considered along with the advantages in the context of specific countries or regions, as a species that is aggressive in one climatic region may be much less so in another (Hitchmough, 2004). It is possible to choose non-invasive exotics (Dunnet, 2010) that could be used for their resistance to salt stress.

4.2 Plant management

The consociation of halophytes with other less salt-tolerant species might enhance the tolerance of the latter if the halophytes, which are salt accumulators, reduce the local salinity through salt bioaccumulation. In an attempt to evaluate this approach, five species (*Suaeda maritima, Sesuvium portulacastrum, Clerodendrum inerme, Ipomoea pes-caprae, Heliotropium curassavicum*) and one tree species (*Excoecaria agallocha*) were used in a trial (Ravindran et al., 2007). Of the six species studied *S. maritima* and *S. portulacastrum* exhibited greater accumulation of salts in their tissues and reduced salts in the soil medium.

Arbuscular mycorrhizal (AM) symbiosis can increase host resistance to drought stress, although the effect is unpredictable. Since water and salt stress are often linked in drying soils, the AM influence on plant drought response can be partially the result of AM influence on salinity stress. With this idea in the mind, Cho et al. (2006) tested the hypothesis that AM-induced effects on drought responses would be more pronounced when plants of comparable size were exposed to drought in salinized soils than when only drought was applied. In two greenhouse experiments, several water relations characteristics were measured in sorghum plants colonized by *Glomus intraradices, Gigaspora margarita* or a mixture of AM species, during a sustained drought following exposure to salinity treatments (NaCl stress, osmotic stress via concentrated macronutrients, or soil leaching). The findings confirmed that AM fungi can alter host response to drought but did not lend much support to the idea that AM-induced salt resistance might help explain why AM plants can be more resilient to drought stress than their non-AM counterparts. As far as we are aware, the value of AM to combating salinity in landscaping has yet to be evaluated.

Shade conditions can also influence the effects of salinity on plants. Devitt et al. (2005b) quantified the foliar damage and flower production of 19 flowering landscape plants, sprinkle irrigated with reuse water and reuse water plus a period of shade (24% reduction in solar radiation). Results indicated that about half of the treated species had an acceptable levels of foliar damage at 1.94 dS m^{-1} in both of the reuse treatments *(Asteriscus maritimus* 'Gold Coin', *Centaurea cineraria, Lantana camara* var. *aculeate, Bougainvillea spectabilis, Gazania* spp., *Hemerocallis fulva, Gaillardia aristata, Mesembryanthemum crystallinum)*. However while the partial shade minimized the negative effect of salt spray application, none of the species tested performed well enough to be used in the landscaping.

The use of enhanced potassium plant nutrition is an efficient method of preventing sodium-induced stress in many crops (Evans & Sorger, 1966) and additional nitrate fertilization can alleviate chloride-induced stress (Bar et al., 1997) so the application of potassium nitrate fertilizer has been shown as a very efficient method of enhancing crop performances under saline conditions (Achilea, 2003). It has to be determined if this procedure is applicable in the landscape, where the maintenance level is low.

5. Irrigation modalities to reduce salt stress

As in agriculture, amenity landscapes that provide ornamental or utility value are irrigated when rain is insufficient to support expected growth. Irrigation to compensate for inadequate rainfall can be permanent in arid areas or temporary when short-term drought threatens. Landscapes have additional irrigation requirements uncommon in agriculture. Most landscape plants need short-term irrigation following planting until they establish new roots in the surrounding soil. Also, plants can be placed in landscape situations of very limited soil-water availability, such as aboveground planters, that require permanent irrigation regardless of the climate (Kjelgren et al., 2000). Sometimes plants that are damaged or dead have to be replaced but this operation is very expensive (e.g. amounting to hundreds of thousands of dollars in a golf course in Southern Nevada, Devitt et al., 2004).

Unlike the monoculture of agricultural crops, most landscape plantings include a variety of species with different abilities to tolerate drought and salt in irrigation water (Wu et al., 2001b) and hence different needs for water. Additionally, the method of irrigation used such as drip, ground surface application or sprinkler irrigation can, as we have seen (Section 3.1), affect the severity of plant damage from salty irrigation water (Miyamoto, 2004; Wu & Dodge, 2005). It has been well documented that applying irrigation water containing soluble salts via drip, bubbler or even flood irrigation has a less damaging effect on plants than applying the same water via overhead irrigation (Benes et al., 1996; Bernstein & Francois, 1975; Gornat et al., 1973; Jordan et al., 2001). Drip irrigation is often preferred, as it also minimizes fluctuations in soil water content (Shalhavet, 1984), maintaining the soil moisture continuously high at the root zone and a low salt concentration. The development of saline bulbs within the root zone can be managed with a sufficient leaching and monitoring of soil EC_e (Boland, 2008). However, even though the majority of plants are more sensitive to sprinkler than drip irrigation, the use of sprinklers is often preferred as it requires less maintenance and is less vulnerable to damage than drip irrigation (Wu & Dodge, 2005). Common problems associated with drip irrigation are the need to remove the accumulated salts from the wetting front and avoidance of drippers clogging. Where flood irrigation is used, the leaching of salts is largely determined by the soil type, as the low hydraulic conductivity of clay soils (especially in the presence of sodicity) often minimizes any opportunity for management or 'strategic leaching' (i.e. leaching at specific times) - although there may be some opportunity for the installation of surface or subsurface drains to assist drainage (Boland, 2008).

Techniques for controlling salinity that require relatively minor changes to schedules are more frequent irrigations, additional leaching, pre-plant irrigation, bed forming and seed placement. Salt concentrations are lowest following an irrigation and highest just before the next irrigation. Increasing irrigation frequency maintains more constant moisture content in the soil so that more of the salts are then kept in solution which aids the leaching process (Fipps, 2003). Applying sufficient volumes of water to allow an adequate salt leaching can alleviate salt accumulation and sustain the prolonged use of alternative water sources for irrigation of landscapes (Niu & Cabrera, 2010). The growth stage can also affect the sensitivity of the plants to salt stress. Woody species demonstrate greater sensitivity to salinity during the early developmental phases, so in this case giving water of better quality is preferred.

The term 'irrigation scheduling' includes both the estimation of the irrigation requirements of the given crop and the appropriate irrigation intervals, which is often complicated to establish in landscapes due to lack of information on the water-use of salt-stressed plants (Paranychianakis & Chartzoulakis, 2005). Precise irrigation scheduling relies on the right amount of water at the right time (Fereres et al., 2003) and a proper choice can have a significant impact on salinity effects through either over or under irrigation. The adoption of irrigation scheduling tools such as monitoring evapotranspiration and soil moisture, the monitoring of salinity (EC in the soil, leaf-sap in the plant), determining adequate leaching or drainage for water-logging-sensitive crops and balancing the accumulation of the salts with leaching irrigations under conditions of limited water supply (Boland, 2008) can assist scheduling decisions.

6. Conclusions

The salt tolerance of landscape plants varies widely with species, environmental conditions and soil or substrate. Landscape plants, most of which are non-halophytes, have similar mechanisms of salt tolerance to agricultural crops, but assessment of salt tolerance for landscape plants should be based primarily on aesthetic value rather than effects on biomass. In relation to the wide number of plant species potentially available, it should be possible choose ornamental genotypes suitable for saline environments. Problems that occur are linked to: 1) the necessity to carry out trials on a wide range of plant species to find those most suitable for specific sites; 2) choosing parameters that are easy to measure to characterize tolerance to salt stress and 3) tailoring irrigation modalities or plant management strategies to enable chosen species to cope with salt stress. The use of saline water can be adopted only where the soil characteristics (e.g. sandy soil) and the nature of the salts present in any irrigation or ground water allow the salt water use without damage to the soil structure. However, there are many places where the use of low quality water should be possible and so improve the quality of landscapes and urban life.

7. References

Achilea, O. (2003). Salinity-induced stress in greenhouse crops is alleviated by potassium nitrate-enhanced nutrition, *Acta Horticulturae* 609: 341-347.

Aktas, H., Abak, K. & Cakmak, I. (2006). Genotypic variation in the response of pepper to salinity, *Scientia Horticulturae* 110: 260-266.

Alarcón, J.J., Sánchez-Blanco, M.J., Bolarín, M.C. & Torrecillas, A. (1993). Water relations and osmotic adjustment in *Lycopersicon esculentum* and *L. pennellii* during short term salt exposure and recovery, *Physiologia Plantarum* 89: 441-447.

Alarcón, J.J., Morales, M.A., Torrecillas, A. & Sánchez-Blanco, M.J. (1999). Growth, water relations and accumulation of organic and inorganic solutes in the halophyte *Limonium latifolium* cv. Avignon and its interspecific hybrid *Limonium caspia* x *Limonium latifolium* cv. Beltlaard during salt stress, *Journal of Plant Physiology* 154: 795-801.

Appleton, B., Huff, R.R. & French, S.C. (1999). Evaluating trees for saltwater spray tolerance for oceanfront sites, *Journal of Arboriculture* 25(4): 205-210.

Azza Mazher, A.M., Fatma El-Quesni, E.M. & Farahat, M.M. (2007). Responses of ornamental plants and woody trees to salinity, *World Journal of Agricultural Science* 3(3): 386-395.

Bañon, S., Fernández, J.A., Ochoa, J. & Sánchez-Blanco, M.J. (2005). Paclobutrazol as an aid to reduce some effects of salt stress in oleander seedlings, *European Journal of Horticulture Science* 70: 43-49.

Bar, Y., Apelbaum, A., Kafkaki, U. & Goren, R. (1997). Relationship between chloride and nitrate and its effect on growth and mineral composition of avocado and citrus plants, *Journal of Plant Nutrition* 20: 715-731.

Benes, S.E., Aragues, R., Grattan, S.R. & Austin, R.B. (1996). Foliar and root absorption of Na and Cl in maize and barley: implications for salt tolerance screening and the use of saline sprinkler irrigation, *Plant and Soil* 180: 75-86.

Bernstein, L., Francois, L.E. & Clark, R.A. (1972). Salt tolerance of ornamental shrubs and ground covers, *Journal of American Society for Horticulture Science* 97: 550-556.

Bernstein, L. & Francois, L.E. (1975). Effects of frequency of sprinkling with saline waters compared with daily drip irrigation, *Agronomy Journal* 67: 185-190.

Blum, A. (1986). Salinity resistance, In: *Plant Breeding for Stress Environments*, A. Blum (Ed.), 1163-1169, CRC Press, Boca Raton.

Boland, A.M. (2008). Management of saline/or recycled water for irrigated horticulture, *Acta Horticulturae* 792: 123-134.

Botequilla Leitão, A. & Ahern, A. (2002). Applying landscape ecological concepts and metrics in sustainable landscape planning, *Landscape and Urban Planning* 59: 65-93.

Boursier, P. & Läuchli, A. (1990). Growth responses and mineral nutrient relations of salt stressed sorghum, *Crop Science* 30: 1226-1233.

Brundtland Report (1987). *The Report of the Brundtland Commission, Our Common Future*, Oxford University Press, 43 pp.

Cabrera, R.I., Rahman, L., Niu, G., McKenney, C. & Mackay, W. (2006). Salinity tolerance in herbaceous perennial, *HortScience* 41: 1054.

Cameron, A., Heins, R. & Carlson, W. (2000). Forcing perennials 102. Firing up perennials. The 2000 edition. *Greenhouse Grower Magazine*, Willoughby, Ohio, and Michigan State University, pp. 7-8.

Cameron, R.W.F., Wilkinson, S., Davies, W.J., Harrison Murray, R.S., Dunstan, D. & Burgess, C. (2004). Regulation of plant growth in container-grown ornamentals through the use of controlled irrigation, *Acta Horticulturae* 630: 305-312.

Cartica, R.J. & Quinn, J.A. (1980). Responses of populations of *Solidago sempervirens* (*Compositae*) to salt spray across a barrier beach, *American Journal of Botany* 67: 1236-1242.

Carvajal, M., Martínez, V. & Cerda, A. (1999). Influence of magnesium and salinity on tomato plants grown in hydroponic culture, *Journal of Plant Nutrition* 22: 177-190.

Cassaniti, C. (2008). *Response of ornamental plants to salinity*, PhD thesis (in Italian).

Cassaniti, C., Leonardi, C. & Flowers, T.J. (2009a). The effect of sodium chloride on ornamental shrubs, *Scientia Horticulturae* 122: 586-593.

Cassaniti, C., Li Rosi, A. & Romano, D. (2009b). Salt tolerance of ornamental shrubs mainly used in the Mediterranean landscape, *Acta Horticulturae* 807: 675-680.

Cassaniti, C. & Romano, D. (2011). The use of halophytes for Mediterranean landscaping, In: *Proceedings of the European COST Action FA0901*, The European Journal of Plant Science and Biotechnology 5 (Special Issue 2), pp. 58-63.

Cheplick, G.P. & Demetri, H. (1999). Impact of saltwater spray and sand deposition on the coastal annual *Triplasis purpurea* (*Poaceae*), *American Journal of Botany* 86: 703-710.

Cho, K., Toler, H., Lee, J., Ownley, B., Stutz, J.C., Moore, J.L. & Augè, R.M. (2006). Mycorrhizal symbiosis and response of sorghum plants to combined drought and salinity stresses, *Journal of Plant Physiology* 163: 517-528.

Choukr-Allah, R. (1997). The potential of salt-tolerant plants for utilization of saline water. *Options Méditerranéennes*, Sér. A/n. 31, Séminaires Méditerranéens.

Clary, J., Savé, R., Biel, C. & De Herralde, F. (2004). Water relations in competitive interactions of Mediterranean grasses and shrubs, *Annals of Applied Botany* 144: 149-155.

Colmer, T.D., Munns, R. & Flowers, T.J. (2005). Improving salt tolerance of wheat and barley: future prospects, *Australian Journal of Experimental Agriculture* 45: 1425-1443.

Cramer, G.R., Epstein, E. & Läuchli, A. (1990). Effects of sodium, potassium and calcium on salt-stressed barley. I. Growth analysis, *Physiologia Plantarum* 80: 83-88.

Cuthbert, R.W. & Hajnosz, A.M. (1999). Setting reclaimed water rates, *Journal of American Water Works Association* 91(8): 50-57.

Darlington, W.A. & Cirulis, N., (1963). Permeability of apricot leaf cuticle, *Plant Phisiology* 38(4): 462-467.

Devitt, D.A., Morris, R.L., Kopec, D. & Henry, M. (2004). Golf course superintendents attitudes and perceptions toward using reuse water for irrigation in the southwestern United States, *HortTechnology* 14: 1-7.

Devitt, D.A., Morris, R.L. & Fenstermaker, L.K. (2005a). Foliar damage, spectral reflectance, and tissue ion concentrations of trees sprinkle irrigated with waters of similar salinity but different chemical composition, *HortScience* 40: 819-826.

Devitt, D.A., Morris, R.L., Fenstermaker, L.K., Baghzouz, M. & Neuman, D.S. (2005b). Foliar damage and flower production of landscape plants sprinkle irrigated with reuse water, *HortScience* 40: 1871-1878.

Dirr, M. (1976). Selection of trees for tolerance to salt damage, *Journal of Arboriculture* 2: 209-216.

Dobson, M.C. (1991). De-icing salt damage to trees and shrubs, *Forestry Commission Bulletin Number 101*, pp. 64.

Dole, M.J. & Wilkins, H.F. (1999). *Floriculture: Principles and Species*, Prentice-Hall, USA, Inc. pp. 613.

Dunnet, N. (2010). People and nature: integrating aesthetics and ecology on accessible green roofs, *Acta Horticulturae* 881: 641-652.

Evans, H.J. & Sorger, J.G. (1966). Role of mineral elements with emphasis on the univalent cations, *Annual Review of Plant Physiology* 17: 47-77.

Fereres, E., Goldhamer, D.A. & Parsons, L.R. (2003). Irrigation water management of horticultural crops, *HortScience* 38: 1036-1042.

Ferrante, A., Trivellini, A., Malorgio, F., Carmassi, G., Vernieri, P. & Serra, G. (2011). Effect of seawater aerosol on leaves of six plant species potentially useful for ornamental purposes in coastal areas, *Scientia Horticulturae* 128: 332-341.

Ferguson, L. & Grattan, S.R. (2005). How salinity damages citrus: osmotic effects and specific ion toxicities, *HortTechnology* 15: 95-99.

Fipps, G. (2003). Irrigation water quality standards and salinity management strategies, *Texas Cooperative Extension*, College Station, TX, Publication Number B-1667.

Flowers, T.J. & Yeo, A.R. (1995). Breeding for salinity resistance in crop plants: where next?, *Australian Journal of Plant Physiology* 22: 875-884.

Flowers, T.J. & Colmer, T.D. (2008). Salinity tolerance in halophytes, *New Phytologist* 179: 945-963.

Forti, M. (1986). Salt tolerant and halophytic plants in Israel, *Reclamation and Revegetation Research* 5: 83-96.

Fox, L.J., Grose, N., Appleton, B.L. & Donohue, S.J. (2005). Evaluation of treated effluent as an irrigation source for landscape plants, *Journal of Environmental Horticulture* 23: 174-178.

Franco, J.A., Fernández, J.A., Bañón, S. & Gonzáles, A. (1997). Relationship between the effects of salinity on seedling leaf area and fruit yield of six muskmelons cultivars, *HortScience* 32: 642-647.

Franco, J.A., Cros, V., Bañón, S. & Martinéz-Sanchéz, J.J. (2002). Nursery irrigation regimes and establishment irrigation affect the postplanting growth of *Limonium cossonianum* in semiarid conditions, *Israel Journal of Plant Science* 50: 25-32.

Franco, J.A., Martinéz-Sanchéz, J.J., Fernández, J.A. & Bañón, S. (2006). Selection and nursery production of ornamental plants for landscaping and xerogardening in semi-arid and environments, *Journal of Horticulture Science and Biotechnology* 81: 3-17.

Francois, L.E. (1982). Salt tolerance of eight ornamental tree species, *Journal of American Society of Horticulture Science* 107: 66-68.

Francois, L.E. & Clark, R.A. (1978). Salt tolerance of ornamental shrubs, trees and iceplant, *Journal of American Society of Horticulture Science* 103: 280-283.

Gori, R., Ferrini, F., Nicese, F.P. & Lubello, C. (2000). Effect of reclaimed wastewater on the growth and nutrient content of three landscape shrubs, *Journal of Environmental Horticulture* 18: 108-114.

Gornat, B., Goldberg, D., Rimon, R. & Ben-Asher, J. (1973). The physiological effect of water quality and method of application on tomato, cucumber and pepper, *Agronomy Journal* 82: 943-946.

Graf, A.B. (1992). *Hortica. Color cyclopedia of garden flora in all climates and exotic plants indoors*, Roehrs Company, East Rutherford, USA, pp. 1216.

Greene, D.W. & Bukovac, M.J. (1974). Stomatal penetration: effect of surfactant and role in foliar absorption, *American Journal of Botany* 61: 100-106.

Hesp, P.A. (1991). Ecological processes and plant adaptations on coastal dunes, *Journal of Arid Environments* 21: 165-191.

Hitchmough, J. (2004). Philosophical and practical challenges to the design and management of plantings in urban greenspace in the 21st century, *Acta Horticulturae* 643: 97-103.

Ibrahim, K.M., Collins, J.C. & Collin, H.A. (1991). Effects of salinity on growth and ionic composition of *Coleus blumei* and *Salvia splendens Journal of Horticulture Science* 66: 215-222.

Iles, J.K. (2003). The science and practice of stress reduction in managed landscapes, *Acta Horticulturae* 618: 117-124.

Johnson, A.M. & Whitwell, T. (1997). Selecting species to develop a field-grown wildflower sod, *HortTechnology* 7: 411-414.

Jordan, L.A., Devitt, D.A., Morris, R.L. & Neuman, D.S. (2001). Foliar damage to ornamental trees sprinkler-irrigated with reuse water, *Irrigation Science* 21: 17-25.

Karakas, B., Lo Bianco, R. & Rieger, M. (2000). Association of marginal leaf scorch with sodium accumulation in salt stressed peach, *HortScience* 35: 83-84.

Kirkam, M.B. (1986). Problems of using wastewater on vegetable crops, *HortScience* 21: 24-27.

Kjelgren, R., Rupp, L. & Kilgren, D. (2000). Water conservation in urban landscapes, *HortScience* 35: 1037-1040.

Kotzen, B. (2004). Plant use in desert climates – looking forward to sustainable planting in the Negev and other world deserts, *Acta Horticulturae* 643: 39-49.

Kozlowski, T.T. (1997). Responses of woody plants to flooding and salinity, *Tree Physiology Monograph N° 1*, Heron Publishing, Victoria, Canada.

Maas, E.V. & Grattan, S.R. (1999). Crop yields as affected by salinity, In: *Agricultural Drainage*, R.W. Skaggs & J. van Schilfgaarde (Eds.), 55–108, Agronomy Monograph 38, ASA,CSSA, SSSA. Madison, WI.

Maas, E.V. & Francois, L.E. (1982). Sprinkler-induced foliar injury to pepper plants: effects of irrigation frequency, duration and water composition, *Irrigation Science* 3: 101-109.

Maathuis, F.J.M. & Amtmann, A. (1999). K+ nutrition and Na+ toxicity: the basis of cellular K+/Na+ ratios, *Annals of Botany* 84: 123-133.

Marosz, A. (2004). Effect of soil salinity on nutrient uptake, growth and decorative value of four ground cover shrubs, *Journal of Plant Nutrition* 27: 977-989.

Marschner, H. (1995). *Mineral nutrition of higher plants*, 2nd Ed., Academic Press, San Diego, California.

Martinéz-Sanchéz, J.J., Ferrandis, P., Trabaud, L., Galindo, R., Franco, J.A. & Herranz, J.M. (2003). Comparative root system structure of post-fire *Pinus halepensis* Mill. and *Cistus monspeliensis* L. samplings, *Plant Ecology* 168: 309-320.

Matsuda, K. & Riazi, A. (1981). Stress-induced osmotic adjustment in growing regions of barley leaves, *Plant Physiology* 68: 571–576.

McCammon, T.A., Marquart-Pyatt, S.T. & Kopp, K.L. (2009). Water-conserving landscapes: an evaluation of homeowner preference, *Journal of Extension* 47(2): 1-10.

Miyamoto, S., Martinez, I., Padilla, M. & Portillo, A. (2004). Landscape plant lists for salt tolerance assessment, *El Paso Water Utilities Public Board Publication*, p. 12.

Morales, M.A., Sánchez-Blanco, M.J., Olmos, E., Torrecillas, A. & Alarçon, J.J. (1998). Changes in growth, leaf water relations and cell ultrastructure in *Argyranthemum coronopifolium* plants under saline conditions, *Journal of Plant Physiology* 153: 174-180.

Morales, M.A., Alarçon, J.J., Torrecillas, A. & Sánchez-Blanco, M.J. (2000). Growth and water relations of *Lotus creticus* plants affected by salinity, *Biologia Plantarum* 43: 413-417.

Munns, R. & Termaat, A. (1986). Whole plant response to salinity, *Australian Journal of Plant Physiology* 13: 143-160.

Munns, R. & James, R.A. (2003). Screening methods for salinity tolerance: a case study with tetraploid wheat, *Plant and Soil* 253: 201-218.

Munns, R. & Tester, M. (2008). Mechanisms of salinity tolerance, *The Annual Review of Plant Biology* 59: 651-681.

Murillo-Amador, B., Troyo-Diéguez, E., García-Hernández, J.L., López-Aguilar, R., Ávila-Serrano, N.Y., Zamora-Salgado, S., Rueda-Puente, E.O. & Kaya, C. (2006), Effect of NaCl salinity in the genotypic variation of cowpea (*Vigna unguiculata*) during early vegetative growth, *Scientia Horticulturae* 108: 423-441.

Musacchi, S., Quartieri, M. & Tagliavini, M. (2006). Pear (*Pyrus communis*) and quince (*Cydonia oblonga*) roots exhibit different ability to prevent sodium and chloride uptake when irrigated with saline water, *European Journal of Agronomy* 24: 268-275.

Navarro, A., Bañón, S., Olmos, E. & Sánchez-Blanco, M.J. (2007). Effects of sodium chloride on water potential components, hydraulic conductivity, gas exchange and leaf ultrastructure of *Arbutus unedo* plants, *Plant Science* 172: 473–480.

Navarro, A., Bañón, S., Conejero, W. & Sánchez-Blanco, M.J. (2008). Ornamental characters, ion accumulation and water status in *Arbutus unedo* seedlings irrigated with saline water and subsequent relief and transplanting, *Environmental and Experimental Botany* 62: 364-370.

Niu, G. & Rodriguez, D.S. (2006a). Relative salt tolerance of five herbaceous perennials, *HortScience* 41: 1493-1497.

Niu, G. & Rodriguez, D.S. (2006b). Relative salt tolerance of selected herbaceous perennials and groundcovers, *Scientia Horticulturae* 110: 352-358.

Niu, G., Rodriguez, D.S. & Aguiniga, L. (2007a). Growth and landscape performance of ten herbaceous species in response to saline water irrigation, *Journal of Environmental Horticulture* 25: 204-210.

Niu, G., Rodriguez, D.S., Aguiniga, L. & Mackay, W. (2007b). Salinity tolerance of *Lupinus havardii* and *Lupinus texenis*, *HortScience* 42: 526-528.

Niu, G. & Cabrera, R.I. (2010). Growth and physiological responses of landscape plants to saline water irrigation: a review, *HortScience* 45: 1605-1609.

Paranychianakis, N.V. & Chartzoulakis, K.S. (2005). Irrigation of Mediterranean crops with saline water: from physiology to management practices, *Agriculture, Ecosystems and Environment* 106: 171-187.

Parnell, J. (1988). Irrigation of landscape ornamentals using reclaimed water, *Proceedings of Florida State Horticultural Society* 101: 107-110.

Percival, G.C. (2005). Identification of foliar salt tolerance of woody perennials using chlorophyll fluorescence, *HortScience* 40: 1892-1897.

Pérez-Alfocea, F., Balibrea, M.E., Alarçon, J.J. & Bolarín, M.C. (2000). Composition of xylem and phloem exudates in relation to the salt tolerance of domestic and wild tomato species, *Journal of Plant Physiology* 156: 367-374.

Pessarakli, M., Marcum, K.B. & Kopec, D.M. (2001). Growth responses of desert saltgrass under salt stress, *Turfgrass and Ornamental Research Report*, University of Arizona, College of Agriculture.

Picchioni, G.A. & Graham, C.J. (2001). Salinity, growth, and ion uptake selectivity of container-grown *Crataegus opaca*, *Scientia Horticulturae* 90: 151-166.

Quist, T.M., Williams, C.F. & Robinson, M.L. (1999). Effects of varying water quality on growth and appearance of landscape plants, *Journal of Environmental Horticulture* 17: 88-91.

Ravindran, K.C., Venkatesan, K., Balakrishnan, V., Chellappan, K.P. & Balasubramanian, T. (2007). Restoration of saline land by halophytes for Indian soils, *Soil Biology & Biochemistry* 39: 2661-2664.

Rodie, S.N. & Streich, A.M. (2009). *Landscape sustainability*, Neb-Guide, University of Nebraska–Lincoln Extension.

Rodriguez, D.S. (2005). *Developing a scientific basics for wastewater application in the Chihuahuan desert*, MS thesis, New Mexico State University, Las Cruces.

Rodríguez, P., Torrecillas, A., Morales, M.A., Ortuño, M.F. & Sánchez-Blanco, M.J. (2005). Effects of NaCl salinity and water stress on growth and water relations of *Asteriscus maritimus* plants, *Environmental and Experimental Botany* 53: 113-123.

Romano, D. (2004). Strategie per migliorare la compatibilità del verde ornamentale con l'ambiente mediterraneo, In: *Il verde in città. La progettazione del verde negli spazi urbani*, A. Pirani (Ed.), 363-404, Edagricole, Bologna.

Ruiz, A., Sammis, T.W., Picchioni, G.A., Mexal, J.C. & Mackay, W.A. (2006). An irrigation scheduling protocol for treated industrial effluent in the Chihuahuan desert, *Journal American Water Works Association* 98: 122-133.

Sánchez-Blanco, M.J., Rodríguez, P., Morales, M.A., Ortuño, M.F. & Torrecillas, A. (2002). Comparative growth and water relations of *Cistus albidus* and *Cistus monspeliensis* plants during water deficit conditions and recovery, *Plant Science* 162: 107-113.

Sánchez-Blanco, M.J., Rodríguez, P., Morales, M.A. & Torrecillas, A. (2003). Contrasting physiological responses of dwarf sea-lavender and marguerite to simulated sea aerosol deposition, *Journal of Environmental Quality* 32: 3338-3344.

Sánchez-Blanco, M.J., Rodríguez, P., Olmos, E., Morales, M. & Torrecillas, A. (2004). Differences in the effects of simulated sea aerosol on water relations, salt content, and leaf ultrastructure of rock-rose plants, *Journal of Environmental Quality* 33: 1369-1375.

Savè, R. (2009). What is stress and how to deal with it in ornamental plants? *Acta Horticulturae* 81: 241-254.

Schönherr, J. & Bauer, H. (1992). Analysis of effects of surfactants on permeability of plant cuticles. In: *Adjuvant and agrichemicals*, C.L. Foy (Ed.), 17-35, CRC, Boca Raton, Florida.

Shalhavet, J. (1984). Management of irrigation with brackish water, In: *Soil salinity under irrigation: Processes and management*, I. Shainberg & J. Shalhavet (Eds.), 298-318, Springer Verlag, New York.

Shannon, M.C. & Grieve, C.M. (1999). Tolerance of vegetable crops to salinity, *Scientia Horticulturae* 78: 5-38.

Singh, K.N. & Chatrath, R. (2001). Salinity tolerance. In: *Application of Physiology in Wheat Breeding*, M.P. Reynalds, I. Ortiz-Monasterio & A. McNab (Eds.), 101-110, CIMMYT, Mexico.

Storey, R. & Walker, R.R. (1999). Citrus and salinity, *Scientia Horticulturae* 78: 39-81.

Stroganov, B.P. (1964). *Physiological basis of salt tolerance of plants*, S. Monson, Jerusalem.

Sykes, M.T. & Wilson, J.B. (1988). An experimental investigation into the response of some New Zealand sand dune species to salt spray, *Annals of Botany* 62: 159–166.

Tester, M. & Davenport, R. (2003). Na+ tolerance and Na+ transport in higher plants, *Annals of Botany* 91: 503-527.

Thayer, R.L. (1976). Visual ecology: revitalizing the esthetics of landscape architecture, *Landscape* 20: 37-43.

Torrecillas, A., Rodríguez, P. & Sánchez-Blanco, M.J. (2003). Comparison of growth, leaf water relations and gas exchange of *Cistus albidus* and *Cistus monspeliensis* plants irrigated with water of different NaCl salinity levels, *Scientia Horticulturae* 97: 353-368.

Townsend, A.M. (1980). Response of selected tree species to sodium chloride, *Journal of Horticulture Science* 105: 878-883.

Townsend, A.M. & Kwolek, W.F. (1987). Relative susceptibility of thirteen pine species to sodium chloride spray, *Journal of Arboriculture* 13: 225-227.

USEPA, 1992. *Manual: guidelines for water reuse*, USEPA, Rep. 625/R-92/004. USEPA, Washington, DC.

Utah Division of Water Resources (2003). *Utah's M&I water conservation plan: investing in the future*. URL: http:// www.water.utah.gov / waterplan /uwrpff/TextOnly.htm

Valdez-Aguilar, L.A., Grieve, C.M., Razak-Mahar, A., McGiffen, M. & Merhaut, D.J. (2011). Growth and ion distribution is affected by irrigation with saline water in selected landscape species grown in two consecutive growing-seasons: spring-summer and fall-winter, *HortScience* 46: 632-642.

Van Arsdel, E.P. (1996). Leaf scorch: detecting the cause, *Arbor Age* 16: 22-28.

Wild, A. (1988). *Russell's soil conditions and plant growth*. 11th edn. Harlow, Longman.

Wu, L., Guo, X., Harivandi, R., Waters, R. & Brown, J. (1999). Study of California native grass and landscape plant species for recycled water irrigation in California landscapes and gardens, *University of California-Davis, Slosson Research Endowment for Ornamental Horticulture Research Report 1998-1999*.

Wu, L., Guo, X. & Harivandi, A. (2001a). Salt tolerance and salt accumulation of landscape plants irrigated by sprinkler and drip irrigation systems, *Journal of Plant Nutrition* 24: 1473-1490.

Wu, L., Guo, X., Hunter, K., Zagory, E., Waters, R. & Brown, J. (2001b). Studies of salt tolerance of landscape plant species and California native grasses for recycled water irrigation, *Slosson Report*.

Wu, L. & Dodge, L. (2005). Special Report for the Elvenia J. Slosson Endowment Fund. URL: http://ucce.ucdavis.edu/files/filelibrary/5505/20091.pdf.

Zollinger, N., Cerny-Koenig, T., Kjelgren, R., Koenig, R. & Kopp, K. (2005). Salinity tolerance of eight ornamental herbaceous perennials, *HortScience* 40: 1034-1035.

Zollinger, N., Koenig, R., Cerny-Koenig, T. & Kjelgren, R. (2007). Relative salinity tolerance of intermountain Western United States native herbaceous perennials, *HortScience* 42: 529-534.

Zurayk, R., Tabbarah, D. & Banbukian, L. (1993). Preliminary studies on the salt tolerance and sodium relations of common ornamental plants, *Journal of Plant Nutrition* 16: 1309-1316.

8

Greywater Use in Irrigation: Characteristics, Advantages and Concerns

Cristina Matos, Ana Sampaio and Isabel Bentes
University of Trás-os-Montes e Alto Douro,
Portugal

1. Introduction

Agriculture and urban irrigation import large volumes of clean water to provide for the water needs. The shortage of freshwater resources is an ever-increasing concern worldwide, particularly in the Middle East and North Africa, where the availability of water is reaching crisis levels and chronic water stress (Jury & Vaux Jr, 2007). The awareness of the natural limitations of this resource is growing and so, water reuse has assumed a great significance. In some countries, like Israel, 70% of the treated wastewater is reused for agriculture irrigation (Mekorot, 2007).

Water resources are being, over decades, intensively over explored and polluted, and it is estimated that in a few years, it is reached highly values of water stress in Europe. Portugal is already in the ranking of countries with medium water stress (10-20%). According to Melo-Baptista, (2002), 87.3% of the volume of water used in Portugal is for agriculture and 91.9% of this volume is inefficiently used which represents 219M€/year.

The amount of water needed for domestic consumption in developed countries is around 100-180 L/hab.dia, representing 30-70% of the amount of water required in an urban area (Friedler, 2004). The increased demand for water leads to demand for new more distant sources and / or greater depths, which leads to increased environmental costs and economic exploitation. Within this context, new approaches are emerging to achieve a more sensible and sustainable management of existing water researches. In fact, to avoid the deterioration of this situation it is imperative to consider different approaches such as water reuse strategies. Indeed, one of ways by which we can reduce the pressure on town water supplies is to reuse greywater for irrigation around household. The use of domestic greywater for irrigation is becoming increasingly common in both developed and developing countries to cope with the water scarcity. The adoption of this and other measures will lead, in Portugal, to the increase of efficiency in the use of water, in agriculture, what will allow savings of 65 M€/year (Melo-Baptista, 2002).

The use of decentralise, alternative water sources such as rainwater or greywater is increasingly promoted worldwide.

2. Greywater reuse

Wastewater reuse in agriculture, after the appropriate treatment, may be a high advantageous technique, once that these wastewaters are very rich in organic matter and nutrients that can be used by the cultures and soils. Wastewater reuse in agriculture, design as "Blackwaters farming" is referenced since the final of XIX century in countries like Australia, France, Germany, India, United Kingdom and USA. In the last 20 years it is observed a growing interest in the use of these wastewaters in irrigation, mainly in the arid and semi-arid regions, where is found a lack of water and a grown need for food production (WHO, 1989).

The water becomes, inside houses, in two types of wastewater, black water and greywater, which is centralized in a single collector mixture towards a system of single treatment. Greywater is defined as the domestic wastewater without the contribution of black water from the toilets, i.e., corresponds to the wastewater from baths, washbasins, bidets, washing machines and dishwashers and kitchen sink (Eriksson et al., 2002). Greywater is usually considered to be high volume with a lower level of pollution while blackwater is low volume with the higher level of pollution (Neal, 1996).

A greywater use system captures this water before it reaches the sewer. Kitchen sink or dishwasher wastewater is not generally collected for use as it has high levels of contamination from detergents, fats and food waste, making filtering and treatment difficult and costly (Matos, 2009). This separation allows creating a light greywater (LGW) for use. So, LGW exclude water from the washing machine, dishwasher and kitchen sink.

Wastewater and greywater recycling are emerging as integral part of water demand management, promoting the preservation of high quality freshwater as well as reducing pollutants in the environment and reducing overall supply costs (Al-Jayyousi, 2003). Recent developments in technology and changes in attitudes towards water reuse suggest that there is potential for greywater reuse in the developing world.

It is estimated that the total amount of greywater corresponds to 50-80% (Hansen & Kjellerup, 1994; Al-Jayyousi, 2003) of the wastewater drained from a house constituting the largest potential source of water saving, if consider the possibility of reuse. Greywater is therefore an important component of wastewater and, qualitatively, studies have shown that there is a significant contribution of this greywater to the concentration of some pollutants and contaminants in the total wastewater. In fact, despite being regarded by many as relatively clean water, greywater can be quite polluted, and its indiscriminate use may represent a risk to public health.

The reuse of greywater *in situ*, may prove to be a practice to consider since its quantity and quality is sufficient to meet the demand for some urban non-potable purposes, such as toilet-flushing, cars-washing and irrigation, since the amount of water required is high and the quality may be lower than the drinking-water.

2.1 Greywater characteristics
2.1.1 Quality parameters

Although conceived to be clean, greywater is polluted and contaminated. Greywater contributes significantly to wastewaters parameters such as biochemical oxygen demand (BOD), chemical oxygen demand (COD), total suspended solids (TSS), ammonium (NH_4^+), total phosphorous, boron, metals, salts, surfactants, synthetic chemicals, oils and greases, xenobiotic substances, and microorganisms (Friedler, 2004; Wiel-Shafran et al., 2006; Travis et al., 2008; Eriksson et al., 2002; Gross et al., 2007; Eriksson & Donner, 2009). Untreated domestic wastewater typically contains 50 to 100 mg/L of oils and greases with approximately 2/3 of the load contributed by greywater (Gray & Becker, 2002;

Tchobanoglous et al., 2003). All of these components have potential negative environmental and health impacts.

There is some research on the quality of greywater and its variation by source (Tables 1 and 2) and within source type (Matos, 2009; Matos et al., 2010). For instance, literature reports important differences for washing machines between the effluents of different cycles and the same can be expected from dishwashers (Rose et al., 1991; Burrows et al., 1991; Christova-Boal et al., 1996; Surendan& Wheatley, 1998; Shin et al., 1998; Nolde, 1999; Eriksson et al., 2002; Friedler, 2004).

Parameters	WC	Bath	Hand-wash	Kitchen sink	Washing machine	Dishwasher
Turbidity (NTU)	60-240[1]	92[2] 28-96[3] 49-69[4]	102*[2]		50-210[1] 108[2] 14-296[3]	
Total solids (mg/L)		631[2] 777-1090[5] 250[6]	558[2] 835[5]	1272-2410[5] 2410[6]	658[2] 350-2091[5] 410-1340[6]	45-2810[5] 1500[6]
Total volatile solids, TVS (mg/L)		318[2] 533[5] 190[6]	240[2] 316*[5]	661-720[5] 1710[6]	330[2] 125-765[5] 180-520[6]	30-1045[5] 870[6]
Total suspended solids, TSS (mg/L)	48-120[1]	76[2] 54-303[5] 120[6] 54-200[7]	40[2] 259[5] 181[7]	625-720[5] 720[6] 185[8]	88-250[1] 68[2] 65-280[5] 120-280[6] 165[7]	15-525[5] 440[6] 235[7]
Volatile suspended solids, VSS (mg/L)		102[5] 85[6] 9-153[7]	36-86[5] 72[7]	459-670[5] 670[6]	97-106[5] 69-170[6] 97[7]	10-424[5] 370[6]
pH	6.4-8.1[1] 7.1[5]	7.6[2] 6.7-7.4[4,5]	8.1[2] 7.0-8.1[5]	6.5[5] 6.3-7.4[8]	9.3-10.0[1] 8.1[2] 7.5-10.0[5]	9.3-10.0[1]
Chemical oxygen demand, COD (mg/L) Total	210-230[5]	645[5] 210-501[7] 100-633[9] 282[10]	95-386[5] 298[7] 383[10]	936[2] 644-1340[5] 1079[7] 1380[10] 15-26[11]	725[2,10] 1339[5] 1815[7]	1296*[5]
Dissolved	165[5]	319[5] 184-221[7]	270[5] 221[7]	679*[5] 644[7]	996[5] 1164[7]	547*[5]
Biochemical oxygen demand, BOD_5 (mg/L) Total	173[5]	216[2] 170[6] 424[5] 192[10]	252[2] 33-236[5] 236[10]	536[2] 1460[6] 530-1450[5] 5[8] 2762[10]	48-290[1] 472[2] 280-470[5] 150-380[6] 282[10]	390-699*[5] 1040[5]

Dissolved	76-200[1] 75[5]	237[5]	93*[5]	377*[5]	48-290[1] 381*[5]	262*[5]
Total organic carbon, TOC (mg/L) Total	91[5]	104[2] 30-120[5] 100[6]	40[2] 119[5]	582*[5] 880[6]	381[5] 100-280[6]	234*[5] 600[6]
Dissolved	47[1]	59[5]	74*[5]	316*[5]	281*[5]	150*[5]
Nitrogen (mg/L) Total	4.6-20[1]	17[6] 5-10[9]		74[6] 15.4- 42.5[8]	1-40[1] 6-21[6]	40[6]
Ammonia	<0.1-15[1] <0.9-1.1[5]	1.6[2] 0.1-0.4[3] 1.2[5] 2[6] 1.1-1.2[7] 1.3[10]	0.5[2] 0.4-1.2[5] 0.3[7] 1.2[10]	4.6[2] 0.6-6.0[5] 6.0[6] 0.3[7] 0.2-23.0[8] 5.4[10]	<0.1-0.9[1] 10.7[2] 0.06-3.5[3] 4.9-11.0[5] 0.4-0.7[6] 2.0[7] 11.3[10]	4.5-5.4[5] 4.5[6]
Nitrates and Nitrites	<0.05- 0.2[1]	0.9[2] 0.4[6] 4.2-6.3[7] 0.4[10]	0.3[2] 6.0[7] 0.3[10]	0.5[2] 5.8[3] 0.3[6] 0.6[10]	0.1-0.3[1] 1.6[2] 0.4-0.6[6] 2.0[7] 1.3[10]	0.3[6]
Phosphorus (mg/L) Total	0.11-1.8[1]	2.0[6] 0.2-0.6[10]		74.0[6]	0.06-42[1] 21-57[6]	68[6]
Phosphates	4.6-5.3[5]	1.6[2] 10-19[5] 1.0[6] 5.3-19.2[7] 0.9[10]	45.5[2] 13-49[5] 13.3[7] 48.8[10]	15.6[2] 13-31[5] 31.0[6] 26.0[7] 0.4-4.7[8] 12.7[10]	101.0[2] 4-170[5] 4-15[6] 21.0[7] 171.0[10]	32-537[5] 32.0[6]

Table 1. Values for physical-chemical parameters and nutrients in greywater.*Mean of 150 samples; [1] Christova-Boal et al. (1996); [2] Surendran& Wheatley (1998); [3] Rose et al. (1991); [4] Burrows et al. (1991); [5] Friedler (2004); [6] Siegrist et al. (1976); [7] Almeida et al. (1999); [8] Shin et al. (1998); [9] Nolde (1999); [10] Laak (1974) ; [11] Hargelius et al. (1995).

Laundry greywater exhibited a high range of the values of suspended solids, salts, nutrients, organic matter and pathogens which arise from washing of clothes using detergents (Christova-Boal et al., 1996). In fact, some activities such as washing faecal contaminated laundry, childcare and showering add faecal contamination to greywater (Ottoson & Stenström, 2003). Occasionally, gastrointestinal bacteria such as *Salmonella* and *Campylobacter* can be introduced by food-handling in the kitchen (Cogan et al., 1999). Greywater may have an elevated load of easily degraded organic material, which may favour growth of enteric bacteria such as faecal indicators and such growth as been reported in wastewater systems (Marville et al., 2001).

Kitchen greywater is reported as the highest contributor of oils and greases in domestic greywater, but oils and greases are present in all greywater streams (Friedler, 2004).

As demonstrated, the chemical, physical and microbiological characteristics of greywater are quite inconstant among households due to the type of detergents used, type of things being washed, life style of occupants and other practise followed at household levels.

Parameters	Kitchen	Laundry	Bathroom
Escherichia coli (number/100 mL)	$1.3 \times 10^5 - 2.5 \times 10^{8(1)}$		
Thermotolerant *E. coli*	$9.4 \times 10^4 - 3.8 \times 10^{8(1)}$		
Faecal Streptococcus	$5.1 \times 10^3 - 5.5 \times 10^{8(1)}$	MPN 23 - < $2.4 \times 10^{3(2)}$ $1 - 1.3 \times 10^{6(3)}$	MPN 79 - $2.4 \times 10^{3(2)}$ $1.0 \times 10^4 - 7.0 \times 10^6$ (3)
Total Coliforms	$6.0 \times 10^4 - 4.0 \times 10^{7(1)}$	MPN $2.3 \times 10^3 - 3.3 \times 10^{5(2)}$ $8.5 \times 10^5 - 8.9 \times 10^{5(3)}$ $7.0 \times 10^{5(4)}$	MPN 500 - $2.4 \times 10^{7(2)}$ $8.2 \times 10^3 - 7.0 \times 10^{4(3)}$ $5.0 \times 10^4 - 6.0 \times 10^{6(4)}$
Faecal Coliforms		MPN 110- $1.09 \times 10^{3(2)}$	MPN 170 - $3.3 \times 10^{3(2)}$ $1.0 \times 10^3 - 2.5 \times 10^{3(3)}$ $6.0 \times 10^2 - 3.2 \times 10^{3(4)}$

Table 2. Range values for microbial parameters analyses in greywaters obtained in kitchen, laundry and bathroom.[1] Günther (2000); [2] Christova-Boal et al. (1996); [3] Siegrist et al. (1976); [4] Surendran&Wheatley (1998). MPN: most probable number .

2.1.2 Quantity parameters

The amount of wastewater generated within a house varies greatly and depends on several factors such as the age and number of occupants, their habits and how they use water. Some European cities can reach to 586 L/ day / fire of wastewater generated. According to NSW (2006) greywater accounts for 68% (Figure 1) of the total wastewater generated mainly composed of baths and showers (49%) and laundry (34%).

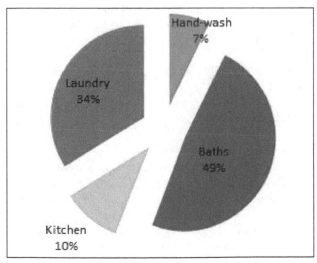

Fig. 1. Distribution (in percentage) of greywater generated in a house (NSW, 2006).

The expression of these quantities in liters per day, based on the reference value to European capitals, has a distribution represented in Figure 2. The differential for the 586 L / day is spent in the toilets.

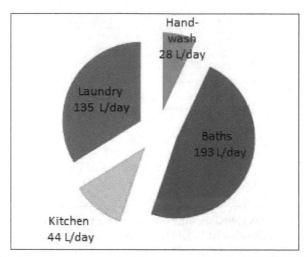

Fig. 2. Distribution of greywater generated in a house in L/day (NSW, 2006).

The capitation varies from country to country. Referring to the example of Israel, Friedler et al. (2005) suggest a capitation from 100 to 150 L/hab.day. In Portugal, it is estimated that each inhabitant spends between 100-180 L/day of water.

Depending on the type of reuse that is considered, all the studies agree on the fact that greywater generated in a house is more than enough to supply inside needs. Friedler, (2004) refers that a greywater reuse scheme would consume only 50-65% of the total greywater produced. Toilet flushing, washing of pavements and cars, and garden irrigation are uses in which the quantity of greywater dispended is high and the needs in terms of quality can be lower than the potable water, and so these can represent potential reuse applications in a unfamiliar dwelling.

Studies that examined the potential of greywater reuse to save freshwater supplies reported savings in the range of 30-50% when greywater is reused for toilet flushing and irrigation (Jeppesen, 1996). When greywater is reused, particularly in garden irrigation considerable volumes of high quality water can be saved.

2.2 Greywater reuse in irrigation

One commonly applied individual initiative to reuse wastewater is the recycling of greywater specifically for irrigation (Travis et al., 2010). In fact, in the past years greywater reuse for irrigation has been considered as a mean of water conservation, since represents the largest potential source of water and costs savings in domestic residence (Al-Jayyousi, 2003), savings up to 38% of water when combined with sensible garden design.

Greywater is a potentially reusable water resource for irrigation of household lawns and gardens (Al-Jayyousi, 2003) as diversion of laundry effluent. According to Jeppesen, (1996) this is technically possible without treatment.

2.2.1 Quality requirements

According to Nolde & Dott (1991), greywater for recycling should accomplish four criteria: hygienic safety, aesthetics, environmental tolerance and technical and economic feasibility. Important parameters to consider for the sustainability of greywater reuse are pH, electrical conductivity, suspended solids, heavy metals, faecal coliforms, *Escherichia coli*, dissolved oxygen, biological and chemical oxygen demands, total nitrogen and total phosphorus (Dixon et al, 1999; Birks & Hills, 2007; Eriksson et al, 2002).

Reuse of greywater for growing plants may affect the microbial activity in the rizosphere that degrades the surfactants and the use by plant for transpiration (Garland et al., 2000). Also, greywater has the potential to increase the soil alkalinity if applied on garden beds over a long time. Greywater with pH values higher than 8 can lead to increase soil pH and reduce availability of some micronutrients for plants.

The various parameter values for the treated wastewater to meet depend on the type of reuse that is proposed. WHO (2006) sets standards in their values of microbiological parameters (Table 4) due to irrigation with wastewater. EPA has already published some guidelines on the reuse of treated domestic wastewater in a variety of purposes, such as agricultural reuse (edible and non-edible crops), urban reuse, and irrigation in areas with restricted access, reuse for recreational purposes, the reuse in construction, environmental reuse, industrial reuse, groundwater recharge and indirect potable reuse.

EPA (2004) classifies agricultural reuse in two subtypes: the reuse by crops not industrially processed and crops industrially processed/non-comestible. In Table 3 there are exposed the quality criteria. The mainly differences relies on admissible BOD and faecal coliforms values, higher in irrigation crops industrially processed.

Parameters	Crops not industrially processed	Crops industrially processed Crops non-comestible
pH	6.9 – 9.0	6.9 – 9.0
BOD (mg/L)	10.0	30.0
Turbidity (NDU)	2.0	n.r.
TSS (mg/L)	n.r.	30.0
Faecal coliform (CFU/100 mL)	Not detectable	< 200
Residual chlorine (mg/L)	1.0	1.0

Table 3. Quality criteria required for agricultural reuse (EPA, 2004). BOD- Biological Oxygen Demand - Standardmethod for indirect measurement of the amount of organic pollution (that can be oxidized biologically) in a sample of water; TSS – Total Suspended Solids - refers to the identical measurement: the dry-weight of particles trapped by a filter, typically of a specified pore size.; n.r. – no reference

WHO divides its criteria in restricted areas of irrigation, that is not accessible, and non-restricted areas. As excepted the criteria are less demanding in the second case.

Parameters	Crops not industrially processed	Crops industrially processed Crops non-comestible
Helminths eggs (n/L)	< 1	< 1
E. coli (CFU/100 mL)	10^5	10^3

Table 4. Microbial quality criteria required for accessible and restricted irrigation areas (WHO, 2006).

According to the NP 4344 concentrations in the wastewater of different elements that constitute a potential risk to the environment should not be higher than the corresponding maximum recommended value (VMR) referred in Decree-law No. 236/98 of 1 August. The physical-chemical parameters referred in Decree-Law as limiting the quality of irrigation water (pH, salinity, sodium absorption ratio, and TSS) should not also exceed the values referenced in Table 5.

	Water quality for irrigation			Water quality for irrigation	
Parameters	RMV	AMV	Parameters	RMV	AMV
Al	5.0	20.0	Mn	0.2	10.0
As	0.10	10.0	Mo	0.005	0.05
Ba	1.0	-	Ni	0.5	2.0
Be	0.5	1.0	NO_3^-	5.0	2.0
B	0.3	0.75	SAR	8.0	-
Cd	0.01	0.05	Salinity (dS/m)	1.0	-
Pb	5.0	20.0	TDS (mg/L)	640	-
Cl-	70.0	-	Se	0.02	0.05
Co	0.05	10.0	TSS (mg/L)	60	-
Cu	0.20	5.0	SO_4^{2-}	575	-
Cr	0.10	20.0	V	0.10	1.0
Sn	2.0	-	Zn	2.0	10.0
Fe	5.0	-	pH	6.5 – 8.4	4.5 – 9.0
F	1.0	15.0	Faecal coliform (CFU/100 mL)	100	-
Li	2.5	5.5	Helminths eggs (n/L)	n.d	1

Table 5. Recommended maximum value (RMV) and Admissible maximum value (AMV) in accordance with Decree-law No. 236/98 (Portugal). Units are in ppm, except when otherwise noted. Sodium Absorption Reason – SAR; Total Dissolved Solids – TDS; Total Suspended Solids – TSS; n.d. – not detectable.

2.2.2 Treatment requirements

Large scale wastewater irrigation programs typically are preceded by conventional treatment measures. However, when wastewater or greywater is reused on the household or in a small property scale, whether due to lack of centralized treatment options or homeowner initiative to save water, adequate treatment is often lacking (Wiel-Shafran et al., 2006).

It is a frequent misconception that greywater is cleaner than combined wastewater and therefore can be reused with minimal or without treatment (Gross et al., 2007). Contrary to public perception, many recent investigations highlight the necessity of greywater treatment before its use on irrigation (Friedler & Gilboa, (2010)).

According to Friedler & Gilboa, (2010), since in on-site systems greywater is reused in close proximity to the general population, safe reuse is possible only after an appropriate treatment that increases its sanitary, environmental and aesthetic quality, which leads to the generally accepted need to provide effective disinfection prior to reuse.

Greywater is often extensively treated in combined systems or separately in spread settlings. The later treatment often consists of a settling tank followed by a soil infiltration system, a sandfilter trench or a subsurface flow wetland providing a reduction of coliforms (Strenström, 1985). The high-grade treatment of greywater has been questioned since it constitutes, as said, a large fraction of the actual wastewater flow, but has a low degree of faecal contamination (Jackson & Ord, 2000) and local systems are often ill adapted for reuse. Al-Jayyousi (2003) described the most common greywater technologies, which divided in Basic two stage systems and biological systems. The first one consists generally in a coarse filtration (thought fibrous of granular depth, or membranes filters) plus disinfection (chlorine or bromine), that employs a short residence time so that the chemical nature of greywater remains unaltered and only minimal treatment is required. The second one involves membrane bioreactors (MBR) and biologically aerated filters (BAF). An alternative approach to disinfection with chlorine is using UV radiation with great results (Friedler & Gilboa, 2010).

According to EPA (2004), the wastewater suitable for irrigation of crops that will not be industrially processed, must go through a secondary treatment, followed by filtration and disinfection. The wastewater suitable for irrigation of crops industrially processed must pass through secondary treatment followed by disinfection.

With regard to the irrigation of non-processed crops, or irrigation of pastures, fields of cereals and other crops not intended for direct consumption, wastewater will have to pass by a secondary treatment, followed by filtration and disinfection, as well as for non-processed crops processed industrially.

2.2.3 Advantages and disadvantages

Below are listed some reported negative effects about greywater irrigation:

- Development of soil hydrophobicity (Chen et al., 2003; Tarchitzky et al., 2007; Wallach et al., 2005);
- Reduction of soil hydraulic conductivity by the surfactants or food-based oils (Travis et al., 2008);
- Surfactants are, as said, a class of synthetic compounds commonly found in greywater and a significant accumulation of these compounds in soils, may ultimately lead to water repellent soils with adverse impacts on agricultural productivity and environmental sustainability (Shafran et al., 2005; Wiel-Shafran et al., 2006);
- Increase of pH in soils and reduced availability of some micronutrients for plants (Christova-Boal et al., 1996);
- Substantial reduction in transpiration rate when pH is above 9 (Eriksson et al., 2006);
- Possibilities of accumulation of sodium and boron in soil, that affects soil properties and plant growth adversely (Misra & Sivongxay, 2009; Gross et al., 2005); Soil aggregate dispersion from sodium accumulation (Misra&Sivongxay, 2009);
- Phytotoxicity due to anionic surfactant content that alters the microbial communities associated with rhizosphere (Eriksson et al., 2006)
- Microbialrisks (Gross et al., 2007);
- Enhanced contamination transport (Grabber et al., 2001);

Sequentially are described some reported positive effects of greywater irrigation:

- Misra et al., (2010) suggested that laundry greywater has a promising potential for reuse as irrigation water to grow tomato, once that compared with tap water irrigated plants, greywater irrigated plants substantially uptake greater quantity of Na (83%) and Fe (86%);
- As said, a large proportion of the ingredients of laundry detergents are essentially non-volatile compounds dominated salts, some of them can be beneficial to plants,

particularly nutrients, although a balanced concentration is required to avoid nutrient deficiency or toxicity in plants (Misra et al., 2010).
• Important water savings and resulting environmental benefits.

2.2.4 Legal aspects

In most countries, until a few years ago, there were no specific guidelines and quality standards for assessing the potential reuse of greywater and associated risks. Legal issues based on alternative related regulations or national discharge limit values, defined for other discharge reuse applications, but not specifically for greywater.

The assessment of water quality until the mid-twentieth century was made based on their aesthetical and organoleptic properties (visual appearance, taste and smell). However, with the progress of science and knowledge, has been coming to the conclusion that this evaluation was insufficient to meet the minimum requirements to protect public health. It has become extremely important to establish normative values for certain parameters that could injury public health.

The World Health Organization (WHO) is a pioneer in defining these values, with the publication of water quality standards, whose first version appeared in the 50's, suffering multiple updates up to today. These standards were the basis for creation in many countries of their own laws. In 1989 the WHO launched a first draft of "Wastewater use in agriculture: guidelines for the use of wastewater excreta and greywater," revised in 2002 and published in 2006. The document, which refers only to the microbiological criteria, should be used for the development of international and national regulations that will assist the management of public health risk associated with the use of wastewater in agriculture and aquaculture.

The development of programs for the use of wastewater began in the twentieth century. The state of California was a pioneer in these programs and appeared in the USA, two statutes that have and continue to have a significant impact on the quantity and quality of wastewater discharged as well as its potential for reuse. These two statutes are called "Water Pollution Control Act" or "Clean Water Act" and "Safe Drinking Water Act". As a result of this law, the centralized WWTP have become common in urban areas, constituting sources of water available for reuse.

The purpose of the "Safe Drinking Water Act" was to ensure that water systems comply with the minimum requirements to protect public health. This allowed the standardization of water quality in the U.S., identifying key contaminants and their maximum limits and indirectly affected the quality of wastewater since the water courses for discharge are often the sources of water supply.

In 1992, the US Environmental Protection Agency (EPA) published "Guidelines for water quality" that describes the treatment stages, water quality requirements and monitoring tools. Later, on 2004, EPA published the "Guidelines for Water Reuse", establishing the nature and extent of treatment and the water quality parameters to impose so that it can be reused. This document also provides some guidelines for monitoring a system for reuse.

The European Union (EU) has published two Directives on the assessment of water quality (Directive 80/778/EEC repealed by Directive 98/83/EC) and Portugal transposed these directives to the internal law, by Decree-Law No. 243/01 of September 5, setting standards for the quality of water for human consumption. Decree-Law 152/97 of 19 June regulates the criteria for collection, treatment and discharge of urban waste water into the aquatic environment. Decree-Law No. 236/98 regulating the quality of water intended for human consumption and is intended to protect public health from the adverse effects of contamination of water.

Directive 91/271/EEC states that the treated wastewater should be reused whenever appropriate and that disposal sites should minimize the adverse environmental effects. The European Commission proposed environmental quality standards that may be used as surrogates for greywater quality assessment in some countries like Portugal (Directive 2000/60/EC, 2006).

In general, the practice precedes the creation of laws. Generally only when there are problems associated with the practice emerge the need for a legal framework. Thus, in Portugal, the legislation directly incident on the field of water reuse is not yet well developed.

The RGAAR approved by Decree No. 23/95 of August 23 addresses the reuse of wastewater very superficially, in particular in Art 11 - Reuse, saying "The treated wastewater and sludge should be reused whenever possible or appropriate."

Marecos do Monte (2008 b), argues that the use of treated wastewater for irrigation, as in Portugal is of great importance, which stems the need for the existence of a standard on this subject that has been published in 2005, the Portuguese Standard NP 4434, "Standard for reuse of treated wastewater for irrigation." This standard represents an important contribution to sustainable practice of reuse of treated wastewater for irrigation, defining:

- Quality requirements for the use of urban treated wastewater as irrigation water;
- The following criteria in the selection of irrigation equipment and processes;
- The procedures to adopt in the implementation of irrigation to ensure the protection of public health and the environment;
- The procedures for the environmental monitoring of the area potentially affected by the irrigation.

The guidelines only applies to the reuse of urban treated wastewater in Wastewater Treatment Plant, in irrigation of agricultural crops, forestry, ornamental ponds, lawns and other green spaces (Marecos do Monte, 2008; Moura et al., 2006).

Despite the normative documents apply to the reuse of urban treated wastewater, these can be used as a basis for guidance on the reuse of treated greywater.

3. Case study: Greywater for irrigation *in situ*

3.1 Introduction

The qualitative and quantitative characterization of the effluent is a key aspect when trying to reuse water. The purpose of this section of the chapter is to characterise, qualitatively and quantitatively, the greywater generated in houses, in order to determine the best treatment and to evaluate the possibility of *in situ* reuse for irrigation.

As it is assumed that the water from the toilets contains high concentrations of contaminants and pollutants, they were eliminated as well as its possibility of reuse. Indeed the aim is to reuse the water by an economically viable process, which would imply the use of untreated wastewater, if possible, or, with a simple/cheap treatment. Therefore it was analyzed the total greywater (TGW), which includes water from all units except the toilet, the light greywater (LGW), that excludes dishwashers, washing machine and kitchen sink from the previous and greywater per domestic device, in order to ascertain what type of water has better features.

With this characterization, it will be possible to outline a feasible reuse strategy using only the greywater of better quality, i.e., excluding the waters from the most polluting sanitary

appliances. It is worth noting say that, the statistical significance of this characterization is limited, since the variability associated with these data is very large (Friedler & Butler, 1996).
In order to reuse it is necessary to know the quality and the quantity of greywater. In fact, to face the possibility of reuse, it is necessary to know the amount of greywater produced by each domestic device.

3.2 Methodology
3.2.1 Greywater characterization: Quality and quantity
3.2.1.1 Total greywater and light greywater quality

In order to characterize total greywater (TGW) produced in households, in the year of 2008, it was changed the drainage system of a dwelling located in Quinta da Casa Nova in Sabrosa, Vila Real District, in Tras-os-Montes and Alto Douro region, northern Portugal (Fig.3). For that purpose, was collect the greywater that came from a bathroom, comprising bath, toilet and bidet, the greywater that came from the kitchen, constituted by the kitchen sink and dishwasher and the greywater from the laundry draining the water generated by the washing machine. The daily occupancy of housing was 4 to 6 people. These wastewaters were sent to a tank in stainless steel AISI 316L, 318 L capacity. The tank capacity was provided in order to collect all the greywater generated during a day, ensuring thus the homogenization of water from various appliances.

Fig. 3. Sabrosa Location.

These are illustrated in Fig.4 and 5.

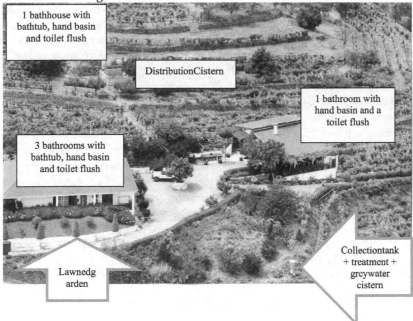

Fig. 4. Quinta da Casa Nova.

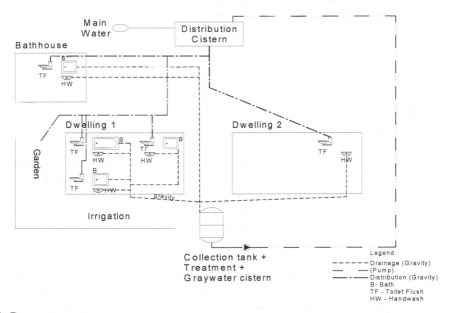

Fig. 5. Greywater system.

Additionally, to characterize light greywater produced in the dwelling it was disconnected from the system described above the drainage of water from the kitchen and laundry. Thus, were drained into the tank previously described only the greywater from the tub, sink and bidet. There have been two sampling campaigns (one in February of 2007 and other in March of 2008) to characterize the TGW and one for the characterization of the LGW (In March of 2008). Also, the potable water physico-chemical characterization was evaluated.

The parameters analysed (Tables 9 and 10) were chosen based on the existing law for irrigation water quality. Given the huge analysis costs, the second campaign was less inclusive, repeating only the most relevant parameters.

In each campaign, it was collected a 5.5 L of greywater sample which was well preserved and sent to a laboratory for the analysis of these parameters. In addition to the above parameters were measured some parameters *in situ* with sensors, such as pH, redox potential, dissolved oxygen and the electrical conductivity. For measuring pH, redox potentialand the electrical conductivity it were used two multisensorial probes, namely a FU20 pH/redox sensor and a ISC40 inductive conductivity sensor, both from YOKOGAWA. To measure dissolved oxygen it was used DO402G-E/U and FD30V27-00-FN/CO5/S50 dissolved oxygen sensor and analyser from YOKOGAWA. This last parameter was only measured for the LGW. The readings of electrical conductivity (Ce) were converted in total dissolved solids (TDS) using the following expression (APHA, 1992):

$$Ce(dS\,/\,m) \times 640 = TDS(mg\,/\,l) \tag{1}$$

The knowledge of the electrical conductivity and TDS allows the evaluation of the water salinity, an important parameter for irrigation reuse.

Knowing sodium, calcium and magnesium content in mg/l (Na^+, Ca^{2+}, Mg^{2+}) it was possible to calculate the sodium absorption reason (SAR):

$$SAR = \frac{Na^+}{\sqrt{\dfrac{Ca^{2+} + Mg^{2+}}{2}}} \tag{2}$$

3.2.1.2 Greywater quality per domestic device

In order to characterize greywater quality per domestic device, independent samples were taken from eight distinct houses collected and treated at the same day. The houses were unifamiliar, varying in the number of inhabitants from 2 to 6 per house. Greywater was separated by its origin and were collected water samples in both rooms that generated effluents: kitchen and bathroom. In each room, waters were collected concerning its origin: (i) in kitchen we took samples in sink, dishwasher and washing machine, and (ii) in the bathroom samples were taken in wash basin, bath and bidet. This last appliance is widespread in Mediterranean Region.

In each sample the following physico-chemical parameters were analysed (*cf. 4.2.1.1.* for probes): pH, electric conductivity, TDS, temperature and COD. All of them, except COD, were analysed with sensors. In respect to microbiological parameters it was determined the total and faecal coliform content in the laboratory, by the membrane filter technique, a highly reproducible method, using standardised selective and solid media (APHA, 1992).

3.2.1.3 Quantitative characterization of greywater produced by

To determine the amount of greywater produced by each domestic device it was performed two sampling campaigns, (7 and 21 days in July of 2008) in 3 different houses, located in Vila Real, Trás-os-Montes and Alto Douro in northern Portugal. During the two campaigns it was observed the volume consumed by each usage, on the counter.

3.2.1.4 Needs for irrigation

To make the quantitative characterization of water demand it was estimated the amount spent in irrigation. To estimate the amount of water spent on irrigation it was conducted a door to door survey in 12 houses with gardens in a residential area of Vila Real, which recorded the number of times per month or per day that there was irrigation and its duration. The consumption data was calculated using the weighted average water consumption of each resident.

3.3 Results and discussion
3.3.1 Greywater characterization: Quality and quantity

3.3.1.1 Total greywater and light greywater quality

The values of the parameters analysed in the TGW, LGW and drinking water are presented in Table 6. In these tables are presented the national legal / regulatory criteria related to water quality for irrigation. Additionally, it presents a range of values, or the average value, depending on the cases, taken from the bibliography. Some of the bibliographic values are presented for greywater from various sources (e.g. kitchen or bathroom) and not necessarily to the mixture of all the greywater.

The most remarkable mark of these waters is the great qualitative variability, which persists even with a high number of repetitions (Friedler & Butler, 1996). In the present study and in agreement with other precedents there were very different values for most parameters, especially with regard to mean concentrations of dissolved oxygen, total coliforms and faecal coliforms.

Given the large range of values indicated in the bibliography, the concentration of most analysed parameters falls within the range of values found by other researchers. It should be noted for the TGW the case of chlorides, and faecal coliform. There were analysed the chlorides while in the bibliography it is presented the total chlorine, which appears with higher concentration values. The value of BOD_5 found is lower than those found in the literature, which is indicative of a lower concentration of organic matter in this sample. For faecal coliform, the value found is higher than the values referenced in the bibliography which could indicate faecal contamination. It is worth noting refer that, in spite of total and faecal coliforms are widely used as indicators of faecal pollution, high levels of them could not necessary indicate pathogen presence (Birks & Hills, 2007), as well their absence does not means that water is pathogen free (Gerba& Rose, 2003). Because some bacterial enteric groups can survive and growth within water closets and pipes (Barker & Bloomfield, 2000), there had been the need to search for more reliable indicators (Scott et al., 2002; Ottosona & Stenström, 2003; Cimenti et al., 2007; Griffin et al., 2008).

With respect to LGW, the parameters values analysed are in the range of values referenced in the bibliography, with the exception of faecal coliform which showed higher values in this campaign and the conductivity that was lower (294 mS/cm) to that presented by Eriksson et al. (2009) (> 700 mS/cm). However, these researchers related this value with the

Parameters	Drinking water	TGW First	TGW Second	LGW	National regulation RMV	National regulation AMV	EPA (2004)	Other References*
Al (mg/L)	0.06	5.8	5.1	1.1	5.0	20.0	-	-
As (mg/L)	<0.01	0.01	-	0.01	0.1	10.0	-	-
Ba (mg/L)	-	0.02	0.02	0.02	1.0	10.0	-	-
B (mg/L)	-	0.2	-	0.2	0.3	-	-	0 - 3.8
Cd (mg/L)	<0.001	0.07	-	0.02	0.01	0.05	-	-
Ca (mg/L)	4.8	9.0	12.0	8.0	-	-	-	-
Pb (mg/L)	<0.005	0.1	-	0.1	5.0	20.0	-	-
Cl- (mg/L)	17.8	72.0	83.0	51.0	70.0	-	-	10.0 [1]
Cu (mg/L)	0.07	0.16	-	0.4	0.2	-	-	-
Cr (mg/L)	<0.002	0.1	-	0.1	0.1	20.0	-	-
Fe (mg/L)	0.02	0.48	0.63	0.93	5.0	-	-	-
Mg (mg/L)	4.8	6.0	7.0	5.0	-	-	-	-
Mn (mg/L)	0.02	0.1	0.1	0.1	0.1	10.0	-	-
Ni (mg/L)	<0.006	0.1	-	0.1	0.5	2.0	-	-
NO_3^- (mg/L)	-	2.0	4.0	2.0	50.0	-	-	0.05 – 74 [2]
Phosphorus(mg/L)	-	8.0	-	2.0	-	-	-	0.1 – 170 [3]
Se (mg/L)	-	0.05	-	0.05	0.02	0.05	-	-
Na (mg/L)	14.8	200.0	170.0	48.0	-	-	-	7.4 - 641
SO_4^{2-} (mg/L)	27.3	130.0	-	14.0	575.0	-	-	-
Zn (mg/L)	-	0.11	0.10	0.22	2.0	10.0	-	0.09 - 6.3
TSS (mg/L)	0.0	51.0	85.0	15.0	60.0	-	-	40 - 120
TDS (mg/L)	46.0	-	-	188.2	640.0	-	-	-
COD (mg/L)	-	720.0	770.0	270.0	150.0	-	-	8000.0
BOD5 (mg/L)	-	170.0	310.0	140.0	40.0	-	≤10	90 - 360
TOC (mg/L)	-	160.0	250.0	1100.0	-	-	-	30 - 880
Total coliform (CFU/100 mL)	n.d.	1.3×10^8	4.8×10^7	4.9×10^6	-	-	-	$70 - 4 \times 10^7$
Faecal coliform (CFU/100 mL)	n.d.	4.3×10^4	3.7×10^3	8.2×10^4	1.0×10^2	-	n.d.	$1 - 9 \times 10^4$
Helminths eggs (n/L)	-	0	0	0	-	-	-	-
Salmonella (CFU/100 mL)	-	0	0	0	1	-	-	-
SAR	-	13.0	51.0	18.8	8.0	-	-	-
pH	6.8	8.9	7.1	6.9	6.5 - 8.4	4.5 - 9.0	6.0 - 9.0	6.4 - 10
Dissolved O_2 (mg/L)	-	7.8	1.3	1.9	-	-	-	2.2 – 5.8
Temperature (ºC)	-	20.0	11.0	16.5	-	-	-	-
Potential Redox (mV)	517.7	-	204.6	164.0	-	-	-	-
Conductivity (µS/cm)	168.0	-	-	294.0	1000.0	-	-	82 - 1565

Table 6. Mean values of the parameters analysed in drinking water, total greywater (TGW) in the first and second campaign, and light greywater (LGW). [1]Total Cl; [2]Total Nitrogen; [3]phosphate; n.d. – Not detectable; Recommended maximum value (RMV) and Admissible maximum value (AMV). * Siegrist et al., (1976); Christova-Boal et al., (1996); Burrows et al., (1991); Rose et al., (1991); Shin et al., (1998); Almeida, et al., (1999); NUEA (2001); Friedler (2004).

high conductivity presented in drinking water from Copenhagen. The same authors argued that the increase of the electrical conductivity is accompanied by an increase in COD, that might indicate the presence of cations as sodium, used in soaps and anions (chloride) used in other types of products such as disinfectants. Also in this work the drinking water conductivity showed a considerable value.

In the second campaign, the amount of dissolved oxygen has been substantially lower than the one of the first campaign, a result consistent with the values obtained for COD and BOD_5, which is higher in this campaign. In fact, the dissolved oxygen decreases or disappears when the water gets large amounts of biodegradable organic substances, since most of the microorganisms responsible for its degradation are aerobic.

As shown by the results presented, LGW still contain large amounts of organic matter and are heavily contaminated (values greater than 104 CFU/100 mL).

Analysing the results from the legal point of view of water reuse for irrigation, it could be argued that the concentration of most parameters in the TGW is not an obstacle. Unlike the aluminium concentration, total suspended solids and chlorides, all above the VMR, and the concentration of cadmium which is above the VMA, limiting the direct use of effluent for this purpose. It should be noted that the value of chlorides of drinking water also was substantial. Also in the LGW, most of the parameters show concentration values that do not limit their application in irrigation. There are, however, some whose concentrations are an obstacle to this application as is the case of faecal coliform, cadmium and copper, whose values are presented above the VMR and selenium with value equal to the VMA. RAS, in this case shows values above the VMR of water for irrigation, thus indicative of a high salinity.

With regard to the microbiological parameters, total and faecal coliforms, LGW were highly contaminated. Consequently, it could not be directly used for irrigation. Considerably decreasing of microbial load could be achieved with sand filtration and coagulation, combined with chorine and UV disinfection (Tajima et al., 2007; Friedler et al., 2008; Friedler & Gilboa, 2010).

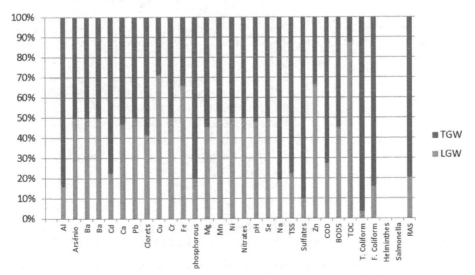

Fig. 6. Relative concentrations of each parameter in Total Greywater (TGW) and Light Greywater (LGW).

In general and as it would be expected, the concentration of the parameters analysed in the TGW is superior to the LGW (Fig. 6). There are, however, some exceptions such as copper (Cu), iron (Fe), zinc Zn) and total organic carbon (TOC), where the concentration is greater in the LGW. For the microbiological parameters, aluminium, cadmium, phosphorus and sodium, TSS, sulphates, COD and RAS concentration difference between the LGW and TGW is evident, and is significantly higher in TGW.

The concentrations values of the parameters is highly variable depending on several factors: since the type of use to the type of detergent used, however, it is most evident pollutant and contaminant load in TGW than in LGW, in particular at the microbiological level, and, in principle, it is easier to treat LGW in order to obtain an effluent for reuse. This finding is in agreement with other referenced work (Almeida et al., 1999; Butler, 1991; Butler et al., 1995)

3.3.1.2 Greywater quality per domestic device

As said before, samples from raw greywater were analysed for pH, conductivity, TDS and COD. In Table 7 there are presented the mean values of each parameter (n=8) by appliance, as well as its standard deviation.

To investigate the concentration of bacteria in raw greywater we enumerated total and faecal coliforms (Table 8).

Source	pH	COD (mgO2/L)	Conductivity (µS/cm)	TDS (mg/L)
Kitchen sink	7.3 ± 0.5	1781.5	150.1 ± 105.8	96.1
Washing machine	10.1 ± 0.3	821.1	3677.1 ± 2826.4	2353.4
Dishwasher	8.5 ± 1.7	1234.5	1560.8 ± 833.8	998.9
Wash-basin	7.1 ± 0.5	196.8	100.9 ±21.1	64.6
Bidet	7.3 ± 0.3	7.9	67.5 ±17.1	43.2
Bath/shower	6.7 ± 1.1	540.2	96.6 ±42.3	60.6
Drinking water	6.7 ± 0.8	-	71.9 ± 73.5	46.0

Table 7. Pollutant concentration per domestic device. COD- Chemical Oxygen Demand; TDS Total Dissolved Solids.

Source	Total coliforms	Faecal coliforms
Wash-basin	$5.4 \times 10^4 \pm 3.5 \times 10^2$	$3.3 \times 10^4 \pm 5.6 \times 10^2$
Bath/shower	$2.2 \times 10^5 \pm 1.1 \times 10^5$	$4.5 \times 10^4 \pm 6.0 \times 10^4$
Bidet	$1.7 \times 10^5 \pm 6.1 \times 10^4$	$2.1 \times 10^2 \pm 3.9 \times 10^2$
Kitchen sink	$6.7 \times 10^6 \pm 3.3 \times 10^5$	$7.0 \times 10^3 \pm 8.9 \times 10^3$
Dishwasher	$2.8 \times 10^6 \pm 2.6 \times 10^5$	$1.5 \times 10^5 \pm 1.7 \times 10^5$
Washing machine	$5.7 \times 10^4 \pm 4.0 \times 10^4$	n.d.
Blended samples	$1.0 \times 10^7 \pm 2.7 \times 10^6$	$2.0 \times 10^5 \pm 6.0 \times 10^4$

Table 8. Total and faecal coliform concentration (CFU/100 mL) for each domestic device (mean of 8 independent samples, with 3 replicas ± standard deviation).n.d. – no detection.

Comparing the mean values of pH recorded for drinking water of different houses with greywater from different sources, it appears that, except for the greywater came from the tub and sink, this value is higher in greywater.

The higher pH values recorded for the water from the washing machines and dishwashers is possibly due to the type of detergents used in the washing. The standard deviation does not assume, in this case, very relevant values. Washing machines and dishwashers reveal again the highest values with respect to conductivity. In fact, the water from the dishwasher has values 20 times higher than the drinking water and water from the washing machine, 50 times higher. The remaining values are close to those recorded for drinking water. The results for this parameter lead to very high SDT values especially in these two domestic devices. It should be noted the high value of standard deviation associated with these results.

The COD values are high, with the exception of water from the bidet, reaching a maximum of 1781.5 mg/L in the sink. Most of the COD derived from the chemicals used and is therefore higher in the laundry and kitchen, with great variations from house to house.

Analysing the results obtained with the purpose of water re-use for irrigation, it could be said that:

- Water for irrigation, requires its improvement and so the separation of sources, distinguishing those which contains a high pH (MLL and MLR). Excluding these waters it is produced a clear greywater with a pH in the range of 6,5-8,4, with features for use in irrigation, under the law (NP 4434, 2005).
- The values of conductivity and TDS present in the MLL and MLR render the direct reuse of water for irrigation, under Decree-Law 236/98, which refers to maximum recommended 1000 mS/cm;
- With regard to microbiological parameters, it make impossible their direct reuse of effluent in irrigation.

Microbiological contamination of total and faecal coliforms is always very significant, with the exception of washing machine that did not presented any faecal coliforms, whatever the dilution used.

Analysing Fig.7 it can be seen that the domestic devices from kitchen and laundry, are the main pollutant concentration producers, although the bath also contained significant amounts of faecal coliform. In fact, the greywater from the kitchen may contain numerous microorganisms from the food washing and is usually the most polluted source.

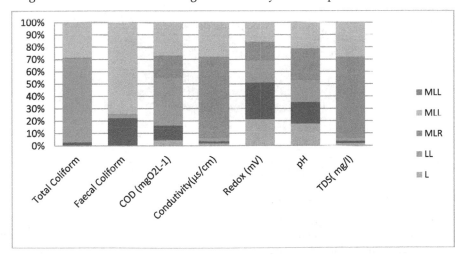

Fig. 7. Relative concentrations of each parameter in each domestic device. MLL- Dish Washer; MLR- Washing machine; LL- Kitchen sink; BA- Bathtub; L- Washbasin;

3.3.1.3 Quantitative characterization of greywater produced by domestic device

In Fig. 8the percentage of water generated by each domestic device is represented.

The capitation found for all sanitary appliances was 114.7 L/person.day, corresponding 95.7 L/person.day to total greywater and 48.6 L/person.day to light greywater.

The study results indicate the great variability associated with the use of some of the sanitary appliances studied, including the MLR, the MLL, the bathtub and the kitchen sink. The high deviations from the average readings for the MLL and MLR are related on the one hand to the fact that the machines were not connected every day and so there were many days of zero consumption. On the other hand, it is related to the type of program used. It should be noted that the sample on the washing machines is not representative, since only one house was equipped with these device. With respect to the tub, the large deviation result on the different habits of the consumers, including the bath duration and the use of water during the same (close or not the tap during soaping).The kitchen sink has a high value of standard deviation, due possibly to the lifestyle of consumers. The fact that consumer's lunch and dinner away can lead to significant deviations from the average. As it has been demonstrated is the bath that is associated with higher value of capitation, followed by the kitchen sink and toilet flushing. The wash basin and the machines occupy a lower share of consumption. These results differ somewhat from those reported in PNUEA, since the latter is associated with 41% of total consumption to flush, followed by 39% to baths and showers. However, the percentage of baths and showers provided by PNUEA,(2005) comes into consideration with the intake valves in the general, without specifying what their origin, and may include a sink and bidet. In this study, washing machines also occupy the lowest-ranking of consumption. Table 9 shows the range of values (maximum and minimum) referenced by Friedler, (2004) concerning the diverse bibliography compiled by this researcher.

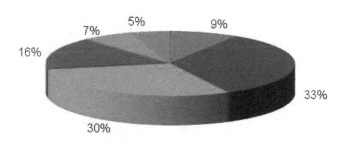

Fig. 8. Percentage of water generated by each domestic device.

Table 9 also shows the limits proposed by the NSW (2006), which can serve as comparison. On this basis we can see that the realities vary greatly. The value of total greywater *per capita* found in this study falls within the range of values that appears in the bibliography.

After made a brief analysis on how much greywater could be expected, it will be interesting knowing if the volume produced is sufficient to meet the demand for *in situ* reuse.

Moreover, knowing the needs, it will be possible to know if it can be reused only the type of greywater less polluted.

Domestic device	Values range	Values mean (n=6)	Friedler (2004)	NSW (2006)
Bath	27.8 – 48.2	38.2	12 - 20	193
Wash basin	7.1 – 12.9	10.4	8 - 15	28
Kitchen sink	17.4 – 50.6	34.0	13 - 25	44
Dishwasher	1.3 – 10.7	6.1	2 - 6	
Washing machine	5.1 – 19.1	7.5	13 - 60	135
Total mixture	48.5 – 141.5	96.2	48 - 126	400

Table 9. Capitation values (L/person.day) found in this study and its comparison with others found in similar studies, in other countries.

3.3.1.4 Needs for irrigation

The maintenance of garden areas and lawns requires a significant amount of water, depending, however, on its geographic location and season. In summer, for example, this volume may represent 60% of the total consumption of a dwelling. Analysing in detail the domestic component can be considered that watering is done only in the 6 months of low rainfall (April to September). In this study it was concluded that the need for irrigation in those months, would be 6794 L/house.month to a garden area of 20 m², implying 226,5 L/house.day (11,5mm/day), one volume, again, easily replaced by greywater, though storage is needed in the months of lower demand. Investigations revealed an average frequency of use in 30 irrigations per month with a duration average of 11.5 minutes per irrigation. According to the PNUEA in the 5 months of lowest rainfall the averages needs of water in a garden located in Portugal are 200 mm/month. According to data from INE (1999), 64% of Portuguese homes are houses, of which 30% have outdoor space and garden or lawn with an average of 40 m²/house. Thus, the average consumption per garden will be 40 m³ per year. According to this plan, in these months, irrigation consumes 266.7 L/house.day (6,7 mm/day), a value lower than the one found in this paper. In Israel, Friedler (2004) states that the reuse of greywater for gardens would need 8-10 L/person.day, or 24,8-30 L/house.day taking in account the average size of the cluster for Portugal. Here it is shown the variability resulting from geographical location and availability of water resources.

Table 10 depicts the amounts of greywater generated by type (supply) and demand for non-potable uses considered.

	Source/use	Volume (L/house/day)
Greywater	TGW	296.7
	LGW	150.7
	Bath	118.4
	Wash basin	31.2
Demand	Irrigation	226.5

Table 10. Amounts of greywater generated by supply type, and demand for non-potable uses.

In conclusion, depending on the type of housing and green areas, the provision of greywater is more than enough to supply the water consumption in toilets, car-washing and to supplement irrigation.

3.4 Conclusions and future recommendations

The results showed that in a reuse perspective it would be best to separate the greywater from the kitchen and laundry of the other sources in order to obtain a clear greywater that would in itself have a better quality. In any case it would have an exempt treatment, even simplified.

Depending on the type of housing and the amount of landscaped green areas, the provision of greywater is enough to supplement the water consumption in irrigation.

There are several possibilities for reuse, which can be considered in order to take full advantage of greywater. The greywater generated in a dwelling, may not be necessary as a whole. Taking into account that the supply is exceeding demand and that the quality of greywater generated can be improved taking into account the separation of sources, it can be assumed the reuse of only part of this water, that is, the one that has the best quality.

4. References

Al-Jayyousi, O.R. (2003). Greywater reuse: towards sustainable water management. *Desalination*, 156, pp: 181-192

Almeida, M.C., Butler, D. & Friedler, E. (1999). At-source domestic wastewater quality. *Urban Water*, 1, pp: 49-55

APHA (1992). Standard Methods for the Examination of Water and Wastewater 18th ed. American Public Health Association, Washington DC In: Greenberg, A.E., Clesceri, L.S., Eaton, A.D.(Eds.)

Barker J.& Bloomfield S. (2000).Survival of *Salmonella* in bathrooms and toilets in domestic homes following salmonellosis. *Journal of Applied Microbiology*, 89, pp: 13-44

Birks, R. & Hills, S. (2007). Characterization of indicator organisms and pathogens in domestic greywater for recycling.*Environmental Monitoring and Assessment*, 129, pp: 61-69

Burrows, W. D., Schmidt M. O.,Carnevale R. M. &Shaub, S. A. (1991). Nonpotable reuse: development of health criteria and technologies for shower water recycle. *Water Science Technology*, 24(9), pp: 81-88

Butler, D., Friedler E. &Gatt K. (1995). Characterizing the quantity and quality of domestic wastewater.*Water Science and Technology*, 31(7), pp: 13-24

Butler, D. (1991). A small-scale study of wastewater discharges from domestic appliances. *Journal of Institution of Water and Environmental Management*, 5 (2), pp. 178–185

Chen, Y., Lerner, O., Shani, U. & Tarchitzky, J. (2003).Hydraulic conductivity and soil hydrophobicity: effect of irrigation with reclaimed wastewater. In: Lundstrum U, editor. Nordic IHSS Symp Abundance and function of natural organic matter species in soil and water, 9th. Sweden: Sundsvall; 2003

Christova-Boal, D., Eden, R.E. & McFarlane, S. (1996). An investigation into greywater reuse for urban residential properties.*Desalination*, 106, pp: 391-397

Cimenti, M, Hubberstey, A., Bewtra, J.K. & Biswas, N. (2007).Alternative methods in tracking sources of microbial contamination in waters.*Water SA*, 33 (2), pp: 183-194

Cogan, T.A., Bloomfield, S.F. & Humphrey, T.J. (1999).The effectiveness of hygiene procedures for prevention of cross contamination from chicken carcasses in the domestic kitchen.*Letters in Applied Microbiology*, 29, pp: 354-358

Decreto-Lei n°236/98 de 1 de Agosto (1998).

Dixon, A.M., Butler, D. & Fewkes, A. (1999). Guidelines for greywater reuse: Health issues. *Journal of the Institution of Water and Environmental Management*, 13, pp: 322-326

EPA (2004).Guidelines for Water Reuse. U.S. Environmental Protection Agency. Washington DC.

Eriksson, E. & Donner, E. (2009). Metals in greywater: sources, presence and removal efficiencies. *Desalination*, 248 (1-3), pp: 271-278

Eriksson, E., Andersen, H.R., Madsen, T.S. & Ledin A. (2009).Greywater pollution variability and loadings.*Ecological Engineering*, 35 (Issue 5), pp: 661-669

Eriksson, E., Auffarth, K., Henze, M. &Ledin A. (2002). Characteristics of grey wastewater.*UrbanWater*, 4 (1), pp: 85-104.

European Comission (2006). Proposal for a Directive of the European Parliament and of the Council on Environmental Quality Standards in the Field of Water Quality Policy and Amending Directive 2000/60/EC, 2006.

Friedler, E., Kovalio, R. & Gail, N.I. (2005). On-site greywater treatment and reuse in multi-storey buildings. *Water science and Technology*, 51(10), pp: 187-194

Friedler, F. & Butler, D. (1996).Quantifying the inherent uncertainty in the quantity and quality of domestic wastewater.*Water Science and Technology*, 33(2), pp: 65-78

Friedler, E. (2004). Quality of individual domestic greywater streams and its implication on on-site treatment and reuse possibilities.*Environmental technology*, 25(9), pp: 997-1008

Friedler, E. & Gilboa, Y. (2010).Performance of UV disinfection and the microbial quality of greywater effluent along a reuse system for toilet flushing.*Science of the Total Environment*, 408, pp: 2109-2117

Friedler, E., Katz, I. &Dosoretz, C.G. (2008).Chlorination and coagulation as pretreatments for greywater desalination.*Desalination*, 222: pp. 38–49

Garland, J.L., Levine, L.H., Yorio, N.C., Adams, J.L. & Cool, K.L. (2000). Greywater processing in recirculation hydroponic systems: phytotoxicity, surfactant degradation and bacterial dynamics. *Water Research*, 34, pp: 3075-3086

Gerba, C.P. & Rose, J.B. (2003). The indicator concept in wastewater reclamation: past, present and future. 4th International Symposium on Wastewayer Reclamation and Reuse, Mexico City.

Gray, S.R. & Becker, N.S.C. (2002). Contaminant flows in urban residential water systems. *Urban Water*, 4 (4), pp: 331-346

Griffin, J.S., Plummer, J.D. & Long, S.C. (2008). Torque teno virus: an improved indicator for viral pathogens in drinking waters.*Virology Journal*, 5: 112 doi:10.1186/1743-422X-5-112

Grisham, A., & Fleming, W.M. (1989).Long-term Options for Municipal Water Conservation. *Journal of the American Water Works Association*, 81 (3), pp: 34-42

Gross, A., Azulai, N., Oron, G., Ronen, Z., Arnold, M. & Nejidat, A. (2005). Environmental impact and health risks associated with greywater irrigation: a case study. *Water Science and Technology*, 52 (8), pp: 161-169

Gross, A., Kaplan, D. & Baker, K. (2007). Removal of chemical and microbiological contaminants from domestic greywater using a recycled vertical flow bioreactor (RVFB). *Ecological Engineering*, 31, pp: 107-114

Günther, F. (2000). Wastewater treatment by greywater separation: Outline for a biologically based greywater purification plant in Sweden. *Ecological Engineering*, 15, pp: 139-146

Hargelius, K., Holmstrand, O. & Karlsson, L. (1995). Hushållsspillvatten Framtagandeavnyaschablonvärdenför BDT-vatten. In Environmental Protection Agency EPA (Ed.), Vadinnehålleravloppfrånhushåll? Näringochmetaller i urinochfekaliersamt i disk-, tvätt-, bad- &duschvatten Stockholm: Swedish EPA (Naturvårdsverket).

Jackson, R. &Ord, E. (2000). Greywater re-use – Benefit or liability? The UK perspective. *Water*, 21: pp: 38-39

Jeppesen, B. (1996). Domestic Greywater reuse: Australian challenge for the future. *Desalination*, 106 (1-3), pp: 311-315

Jury, W.A., Vaux Jr., H.J. (2007). The emerging global water crisis: managing scarcity and conflict between water users. *Advances in Agronomy*, 95, pp: 1-76

Laak, (1974). Relative pollution strengths of undiluted waste materials discharged in households and the dilution waters used for each Manual of grey water treatment practice (pp. 68-78). Michigan, USA: AnnArbor.

Manville, D.,Kleintop, E., Miller, B., Davis, E., Mathewson, J. & Downs, T. (2001). Significance of indicator bacteria in a regionalized wastewater treatment plant and receiving waters. *International Journal of Environment and Pollution*; 15 (4), pp: 461-466

Marecos do Monte, M.H.F. (2008). Sustainable Water Reuse in Portugal. *WSEAS Transactions of Environment and development*, 4 (9), pp: 716-725

Marecos do Monte, M.H.F. (2008b). Portugal implements sustainable water reuse. *Proceedings of 4th IASME/WSEAS International Conference on ENERGY, ENVIRONMENT, ECOSYSTEMS and SUSTAINABLE DEVELOPMENT (EEESD'08)*, pp. 478-483. ISBN 978 960 6766 71 8, Algarve, Portugal, June 11-13, 2008

Matos, C. (2009). Reutilização de água: Utilização de águas cinzentas *insitu*. PhD. Thesis. Universidade de Trás-os-Montes e Alto Douro. Pp: 167.

Matos, C., Sampaio, A. & Bentes, I. (2010). Possibilities of greywater reuse in non-potable in situ urban applications, according with its quality and quantity. *WSEAS Transactions on Environment and Development 7* (6), pp: 499-508

Mekorot. Wastewater Treatment and Reclamation.*Watec 2007*, TelAviv , Israel ; 2007

Melo-Batista, J. (2002). A melhoria da eficiência do uso eficiente da água como contributo para a sustentabilidade dos recursos naturais 10 Encontro Nacional de Saneamento Básico: Uso sustentável da água: situação portuguesa e perspectivas de futuro.

Misra, R.K., Patel, J.H. &Baxi, V.R. (2010). Reuse potential of laundry greywater for irrigation based on growth, water and nutrient use in tomato. *Journal of Hydrology*, 2010, 386 (1-4), pp: 95-102

Misra, R.K. &Sivongxay, A. (2009). Reuse of laundry greywater as affected by its interaction with saturated soil. *Journal of Hydrology*, 366, pp: 55-61

Moura, B., Dionísio, L., Beltrão, J. & Borrego, J.J. (2006). Reclaimed wastewater for golf course irrigation.*WSEAS Transactions on Environment and Development* 2 (5), pp: 652-658

Neal, J. (1996). Waste water reuse studies and trial in Canberra. *Desalination*, 106, pp: 399-405

Nolde, E. (1999). Greywater reuse system for toilet flushing in multi-storey buildings- over ten years experience in Berlin. *Urban water*, 1, pp: 275-284

Nolde, E. &Dott, W. (1991).Verhalten von hygienischbakterien und Grauwasser- Einfluss der UV-Desinfektion and Wiederverkeimung.*GwfWasserAbwasser*, 132 (3), pp: 108-114

Norma Portuguesa 4434- Reutilização de águas residuais urbanas tratadas para rega (2005).

NSW (2006).Guideline for sewered Residential Premises (Single households) Greywater Reuse.

Ottosona, J. &Stenström, T.A. (2003).Faecal contamination of greywater and associated microbial risks.*Water Research*, 37, pp: 645–655

PNUEA (2001). Programa Nacional para o Uso Eficiente da Água. MAOT-IA Lisboa.

Rose, J. B., Sun, G-S., Gerba, C.P. & Sinclair, N.A. (1991). Microbial quality and persistence of enteric pathogens in graywater from various households sources. *Water Research*, 25 (1), pp: 37-42

Scott, T.M., Rose, J.B., Jenkins, T.M., Farrah, S.R. & Lukasik, J. (2002). Microbial Source Tracking: Current Methodology and Future Directions. *Applied and Environmental Microbiology*, 68 (12): 5796–5803

Shin, H.-S., Lee, S.-M., Seo, I.-S., Kim, G.-O., Lim, K.-H.& Song, J.-S. (1998). Pilot scale SBR and MF operation for the removal of organic and nitrogen compounds from greywater. *Water Science Technology*, 38 (6), pp. 79-88

Siegrist, H., Witt M. & Boyle, W.C. (1976).Characteristics of rural household wastewater. *Journal of the Environmental Engineering Division*, 102(EE3), pp: 533-548

Stenström, T. (1985). Infiltration i mark. (Infiltration into soil). Swedish EPA, snv pm 3051.

Surendran, S., Wheatley, A. D. (1998).Grey-water reclamation for non potable reuse.*Journal of the Chartered Institution of Water and Environmental Management*, 6, pp: 406-413

Tarchitzky J., Lerner, O., Shani, U., Arye, G., Lowengart-Aycicegi, A., Brener, A. & Chen, Y. (2007). Water distribution pattern in treated wastewater irrigated soils: hydrophobicity effect. *European Journal of Soil Science*, 58 (3), pp: 573-588

Tchobanoglous, G., Burton, F.L. &Stensel, H.D. (2003).*Wastewater Engineering: treatment and reuse*. New York: Mcraw-Hill, 1819 pp.

Travis, M.J., Weisbrod, N. & Gross, A. (2008). Accumulation of oil and grease in soils irrigated with greywater and their potencial role in soil water repellency.*Science of the Total Environment*, 394 (1), pp: 68-74

Travis, M.J., Wiel-Shafran, A., Weisbrod, N., Adar, E. & Goss, A. (2010).Greywater reuse for irrigation: effect on soil properties. *Science of the Total Environment*, 408 (12), pp: 2501-2508

Wallach, R., Ben-Arie, O. & Graber, E.R. (2005).Soil water repellency induced by long-term irrigation with treated sewage effluent. *Journal of Environmental Quality*, 34 (5), pp: 1910-1920

WHO.(1989). Health guidelines for the use of wastewater in agriculture and aquaculture. WHO, 1989, Technical Report Series, No. 776.

WHO.(2006). Guidelines for the Safe Use of wastewater, excreta and Greywater. Volume I: Policy and Regulatory Aspects. WHO, UNEP, 2006.

Wiel-Shafran, A., Ronen, Z., Weisbrod, N., Adar, E. & Gross, A. (2006). Potential changes in soil properties following irrigation with surfactant-rich greywater. *Ecological Engineering*, 26 (4), pp: 348-354

9

Water Regime Affecting the Soil and Plant Nitrogen Availability

Adrijana Filipović

Faculty of Agriculture and Food-Technology,
University of Mostar,
Bosnia and Herzegovina

1. Introduction

Nitrogen is a necessary ingredient in soil for agriculturalists to produce high-yielding crops. Europe is one of the world's largest and most productive suppliers of food and fiber (Olesen & BIndi, 2002).These authors provide information that in 2004 Europe produced 21% of global meat production and 20% of global cereal production. About 80% of these global productions have occurred in Europe, defined here as the 25 European countries, EU25 (IPCC, 2007).The productivity of European agriculture is generally high, in particular in Western Europe: average cereal yields in the EU are more than 60% higher than the global average (EFMA, 2010). Some plants (legumes in appropriate conditions) produce their own nitrogen (Dorn, 2011) and some nitrogen is contributed to the soil by rainfall, but these natural sources of nitrogen do not occur in high enough levels for prolific crop production. Many agriculturalists add nitrogen to the soil without regarding the plant needs or nutrient soil status. Addition of nitrogen to the soil helps in the rapid and healthy growth of the plants and thus improves the yields of the crops. It also increases the protein content in the crops as well and food value of crop. However, when N inputs to the soil system exceed crop needs, there is a possibility that excessive amounts of nitrate (NO_3^-) may enter either ground or surface water (O'Leary et al., 2002). Managing N inputs to achieve a balance between profitable crop production and environmentally tolerable levels of NO_3^- in water supplies should be every grower's goal. A recent estimate of the current human population supported by synthetic fertilizer is 48%, 100 years after the invention of the synthesis of ammonia from its elements (Erisman et al., 2008). To maximize crop production, the availability of cheap fertilizer in the industrialized world led to excessive use of nitrogen, resulting in a large nitrogen surplus and increased nitrogen losses. The behavior of N in the soil system is complex, yet an understanding of basic processes (mineralization) is essential for a more efficient N management program. Nitrogen, present or added to the soil is subject to several changes (transformations) that dictate the availability of N to plants and influence the potential movement of NO_3^- to water supplies. Nitrogen can be lost from the soil system in several ways: leaching, denitrification, volatilization, crop removal, soil erosion and runoff. And these ways of N losses from agriculture or industry through the global environment system can cause a numbers of different environmental effects: loss of biodiversity, eutrophication of waters and soils, drinking water pollution, acidify cation, greenhouse gas emissions, human health risks through exposure to oxidized nitrogen (NO_x), ozone (O_3) and particulates, and destruction of the ozone layer. Nitrogen fertilizer data throughout the world shows that the annual use rate is increasing (Davidson 2009; FAO, 2010). There is no way to totally prevent the movement of some nitrogen forms to water supplies, but sound management practices can keep losses within acceptable limits. Most of the country was developed Nitrogen Fertilizer Management Plan with the purpose of managing N inputs for crop production to prevent degradation of water resources while maintaining farm profitability. The central tool for achievement of this goal is the adoption of Best Management Practices for Nitrogen. Best management practices for N are broadly defined as economically sound, voluntary practices that are capable of minimizing nutrient contamination of surface and groundwater. The primary focus of the BMP's is commercial N fertilizers; however, consideration of other N sources and their associated agronomic practices is necessary for effective total N management. One of these practices will be presented in further text as usage of diagnostic tool for detecting a nutrient status in plant. We will present recent work on different varieties of potato crop fertilized with increasing nitrogen rate. Rapid methods Cardy ion meter and Chlorophyll meter were tested in open field To collect readings as reaction on nutrient regarding the environmental conditions of cultivars growth and in comparison with standard methods. An effective plant nutrient management practice optimizes nitrogen (N) use efficiency for minimized environmental impact, while ensuring an optimum N status of the crop for good product quality and maximum growth. Soil or plant analysis can be used to evaluate the practice; however the use of plant analysis for this purpose has been limited. One reason is lack of reliable reference values for the critical concentration needed for optimal growth and the other is susceptibility of tools on environmental conditions. Nutrients used for plant growth and biomass productions generally come from the internal cycling of reserve materials which require water for their solubility and translocation, so it can be very variable.

2. Crop yield and quality affected by fertilizer use

Crop response to applied N and use efficiency are important criteria for evaluating crop N requirements for maximum economic yield. Recovery of N in crop plants is usually less than 50% worldwide (Fageria & Baligar, 2005). Low recovery of N in annual crop is associated with its loss by volatilization, leaching, surface runoff, denitrification and plant canopy. Low recovery of N is not only responsible for higher cost of crop production, but also for environmental pollution. Hence, improving N use efficiency (NUE) is desirable to improve crop yields and quality, reducing cost of production and maintaining environmental quality. To improve N efficiency in agriculture, integrated N management strategies that take into consideration improved fertilizer along with soil and crop management practices are necessary. Synchrony of N supply with crop demand is essential in order to ensure adequate quantity of uptake and utilization for optimum yield. Practice of reducing NO_3 loss through soil-plant system include improved timing of N application at appropriate rates, using soil tests and plant monitoring, diversifying crop rotations, using cover crops, reducing tillage, optimizing N application techniques, and using nitrification inhibitors. Today many surveys are focused on understanding methods to minimize NO_3 contamination of water resources and professionals put lot of effort to educate the public about the complexity of the problem and the need for multiple N management practice to solve the problem across agricultural landscapes. The results in the text showed application of tools to monitor the N status of the aboveground canopy of potato, such as chlorophyll readings, sap NO_3-N concentrations, N indices to understand plant's demands and status of nutrient as small step to enhance environmental quality and improvement of product quality for the benefit of producers, processors and consumers. For some crops there has to be a balance with maximum yields and quality. Although N deficiencies will decrease yield, excessive N applications can affect the quality of grains, tubers, fruits and other cropping systems. At higher than needed N levels, quality of malting barley (*Hordeum vulgare* L.) can have undesirable high levels of proteins (Zubrinski et al., 1970; Bishop & MacEachern, 1971). Excessive applications of N fertilizer can decrease tuber quality of potatoes (*Solanum tuberosum* L.) (Laughlin, 1971; Painter et al., 1977; Westermann & Kleinkopf, 1985; Errebhi et al., 1998) and sugar beets (Hills & Ulrich, 1971; Cole et al., 1976; Carter & Traveller, 1981; Hill, 1984). Fruit quality can also be affected by high N rates (Locascio et al., 1984). When quality is an important factor in economic returns such as maximizing production, best management practices that can supply and maintain appreciable N levels for maximum yields and quality are needed. These include practices that can provide high N supply during periods of maximum demand and not to supply excessive N that may decrease product quality. Plant analysis has been considered a very promising tool to assess nutritional requirements of plants. Plant analysis, in conjunction with soil testing, becomes a highly useful tool not only in diagnosing the nutritional status but also an aid in management decisions for improving the crop nutrition (Rashid, 2005). Whatever tool is used, the aim is to serve as an indicator of the actual nutrient status of the soil-plant system. Indicators can be used to evaluate the actual plant nutrient management practice (diagnostic indicators) or to give predictive information such as information on the actual fertilizer requirement for the next application (prognostic indicators) (Lewis, 1993; Schröder et al., 2000). The use of indicators to evaluate the actual practice implies a participatory learning process by which the farmer's motivation for a change is encouraged (Röling & Wagemakers, 2000). The management of plant nutrients can be successively improved by evaluation of the fertilizer practice. Generally, an ideal indicator must be reproducible (Schröder et al., 2000). For evaluation of the nutrient status, the indicator should interpret the actual nutrient status of the soil-plant system in the same manner over different sites and years. Plant analysis are widely used for identifying plant nutrition deficiencies and disturbances in crops but only to a lesser degree for routine evaluation of the plant nutrient status for adequate plant nutrient management. An evaluation of the nutrient status is made possible only by relating the actual status to a standard (Ekbladh, 2007). The material of living plants were consisted of organic matter, water and minerals. The relative amounts of these tree components may vary, but for green plant material, water is always present in the highest proportion and the minerals in the lowest. The percentage distribution of these three components is in the following order of magnitude: water 700 mg g^{-1} fresh matter, organic material 270 mg g^{-1} fresh matter and minerals 30 mg g^{-1} fresh matter. The minerals makes only a comparatively small proportion of the dry matter. They are nevertheless of extreme importance because they enable the plant to build up organic material (photosynthesis). But the ratios of these three components of plant material are highly dependent on environmental conditions.

3. Nitrogen affecting the water quality

The enrichment of nitrogen in the aquatic system impairs the water quality of rivers, lakes, aquifers and coastal and marine waters contributing to the phenomenon of eutrophication (European Environment Agency, 2001). Groundwater is an important resource in Europe, providing water for domestic use for about two third of the population but groundwater is a finite and slowly renewed resource and over exploitation associated with a degradation of water quality is putting in danger an important source of drinking water. In Europe, groundwater nitrate concentrations have remained stable and high in some regions (European Environment Agency, 2005). Most of the nitrates found in groundwater are thus of anthropogenic origin and mostly related to agricultural activities. Indeed nitrogen surplus in agricultural land can be removed by surface runoff, leaching to the aquifer, and loss to the atmosphere or can be stored in the soil–water system. Nitrogen surplus from agriculture are still high in many countries and huge quantities of nitrogen are stored in the soil or aquifers (Grizzetti et al., 2008). There are some major concerns as the Eastern countries will probably intensify their agriculture and thus their fertilization rate in the near future and some countries of Western Europe have not seen their nitrogen surplus decrease but rather stabilize at high levels. Efforts have been taken through conventions or the application of binding Directives and still Europe's waters are suffering from excess nitrogen. Currently in many countries there are strict limits on the permissible concentration of nitrate in drinking water and in many surface waters. The limit is 50 mg NO_3/l in the European Drinking Water Directive (Directive 98/83/EC) and 44 mg NO_3/l in the United States (equivalent to 11.3 mg N/l and 10 mg N/l, respectively). These limits are in agreement with WHO recommendations established in 1970 and recently reviewed and reconfirmed (WHO, 2007; the exact formulation of the standard is that the sum of $NO_3/50$ + $NO_2/3$ should not exceed 1). The European Nitrates Directive also sets a limit concentration of 50 mg NO_3/l for groundwater and surface water as a threshold value for Member States to protect water bodies. Today the agriculture is identified as the single largest source of impairments for water sources. Nitrogen is one of the most abundant elements. About 80 percent of the air we breathe is nitrogen. It is found in the cells of all living things and is a major component of proteins. Inorganic nitrogen may exist in the free form as a gas N_2 or as nitrate NO_3^-, nitrite NO_2^- or ammonia NH_3^+. The rate, time and method of nitrogen application can affect the risk of nitrogen loss to surface water and groundwater. Leaching of nitrate to groundwater and nitrate in subsurface drainage is typically more concentrated with higher nitrogen rates, but the effect of nitrogen rate varies across locations. Soil nitrogen levels and crop needs often are not defined by field borders. Variable types N fertilizers allow farmers to apply fertilizer when and as needed, thus reducing nitrogen loss of water resources, reducing loss of nitrogen to water resources. Achieving a balance of productivity, profit and water quality protection is the goal for nitrogen rate optimization. Not only does the nitrogen impact the water quality, But also have effect on human health. There are two main health issues related to nitrate in drinking water: the linkage with infant methaemoglobinaemia, also known as blue baby syndrome and with cancers, for example of the digestive tract (Ward et al., 2005).

4. Plant analysis as a diagnostic tool for plants nutrition disorder

Plant tissue analysis shows the nutrient status of plants at the time of sampling. This, in turn, shows whether soil nutrient supplies are adequate. In addition, plant tissue analysis will detect unseen deficiencies and may confirm visual symptoms of deficiencies. Toxic levels also may be detected. Though usually used as a diagnostic tool for future correction of nutrient problems, plant tissue analysis from young plants will allow a corrective fertilizer application that same season. Using established critical or standard values, or sufficiency range, a comparison is made between the laboratory analysis results with one or more of these known values or ranges in order to access the plant's nutritional status (Jones et al., 1991; Kelling et al., 2000; Rashid, 2005). The use of plant analysis as a diagnostic tool has a history dating back to studies of plant ash content in the early 1800's. While working on the composition of plant ash, researchers recognized the existing relationships between yield and the nutrient concentrations in plant tissues. Quantitative methods for interpreting these relationships in a manner that could be used for assessing plant nutrient status arose from the work of Macy (1936). Since then, much effort has been directed towards plant analysis as diagnostic tool. Plant analysis is carried out as a series of steps that include sampling and sample preparation followed by laboratory analysis and interpretation of analytical data. Each step is equally important to the success of the technique employed for diagnosing nutritional disorders. Since plant species, plant age, plant part, sampling time and applied fertilizer are all variables that affect the interpretation of the analytical data; careful sampling is highly important (Jones et al., 1991). Surveys of nutrient concentrations in "deficient" and "adequate" N rate for potato crop have been used to establish standard nutrient concentrations. Vegetation period should also be taken in consideration. Nutrient deficiencies are often difficult to identify because a number of interacting factors may cause similar symptoms. Factors such as pests, unfavorable placement, soil chemical properties, soil compaction, or moisture stress can prevent nutrient uptake even if nutrients are plentiful in the soil. Plant tissue analysis will indicate if the crop took up soil-applied nutrients. For perennial fruit crops (blueberries, strawberries, apples, grapes, peach, etc.), these analysis are the best way to monitor the plant's nutrient needs. Plant analysis can be used to fine tune the efficiency of a fertilizer program before nutrient deficiency symptoms occur and is very useful in improving the fruit quality and yield. From emergence through the first few weeks of growth depending on phenophases of sampling crop, plant analysis is helpful in identifying nutrient uptake. Testing the leaf samples at this early stage may indicate where additional nitrogen should be applied. This is a way to determine the cost-effectiveness of the additional application of fertilizer. In field research on different location we have tested two diagnostic tools in interpretation nitrogen status in three varieties of potato crop. Different location were used to compare variation in values depending on micro-climes condition of growth even the same fertilization treatments were applied. Varieties with different vegetation period were also compared to evaluate the distinct in nutrient accumulation rate. For measurement of nutrient status it have been used Chlorophyll meter for evaluation of chlorophyll index in potato leaf and Cardy ion meter for evaluation of nitrate-nitrogen concentration in petiole sap of plant. Data from both measurements were compared to the laboratory analysed leaf on total nitrogen concentration.

4.1 Rapid nitrogen diagnostic tools
Potato plants deficient in N have pale green leaves, poor growth and reduced yield. Excess N fertilizer application increases the chances of surface and ground water contamination. Many farmers in developed countries use a pre-season soil nitrate test to adjust N fertilizer rates to specific potato yield goals. Use of new instruments that instantly estimate potato plant N levels may allow farmers to precisely target N fertilizer applications to changing weather and

crop conditions during the growing season. Fertilizer N utilization efficiency increased when N is applied near the time of greatest need of the crop. Beside basic fertilization application for potato it is common additional treatment usually 45 day after the planting. We have used in our survey two meters that estimate plant N Chlorophyll Meter –(Soil Plant Analysis Development-SPAD 502 meter, Minolta, Osaka,. Japan) (Figure 1) and Cardy-ion Meter (Figure 2). Values obtained by these meters were compared to the standard laboratory measurement of total leaf N expressed on dry weight basis (%) by Kjeldahl method (AOAC, 1970).

The Chlorophyll Meter sensor clamps on intact leaves and instantly measures leaf chlorophyll "greenness". Because there is a close relationship between chlorophyll level and leaf N the Chlorophyll Meter readings (SPAD values) are an indicator of leaf N level (Spectrum technologies, 2011). It is important to take the reading on about the same location on each leaf. It works well to collect the reading from a point one-half the distance from the leaf tip to the collar and halfway between the leaf margin or edge and the leaf midrib. Chlorophyll meter readings are usually stable during the day unless plants are under water stress. As long as readings are collected from the reference strip and the adjacent bulk field at about the same time, the comparison is valid. It is best to avoid collecting readings when moisture is on the leaves (i.e., after a rain or sprinkler irrigation or in the early morning) or when plants are under drought stress as this can distort the readings. Meters should not be subjected to extreme temperature changes before making measurements. Although the chlorophyll meter enables user to quickly and easily measure leaf greenness which is affected by leaf chlorophyll content, several other factor affect SPAD values. Differences in leaf thickness reflected in specific leaf weight are largely responsible for variations in the relationship between N content and SPAD values (Peng et al., 1993). Moreover, the linear relationship of SPAD values and N status in crops varies, depending on growth stages and cultivars. Finally, environmental and stress factor caused by excess or limited water, deficiency of nutrient other then N and pest and diseases can also confound the SPAD readings (Smeal and Zhang, 1994). Producers should recognize this as another tool that may complement, but does not replace, other aspects of sound N management. One soil scientist said it succinctly: "Use the chlorophyll meter to schedule your last 50 lbs N/acre, not your first." Because it is suggested that at least one-half to three-quarters of the total fertilizer N should be applied to the entire field prior to the stage of three leaf to ensure the chlorophyll meter technique effective.

The Cardy ion meter has a sensor that measures the nitrate concentration in liquid extracts of plant tissues. Nitrate moves from roots to leaves in potato plant where were assimilated into amino acids and proteins. While plants take up nitrogen in both the ammonium and nitrate forms, nitrate is usually more abundant than ammonium so nitrogen tests measure nitrate rather than ammonium. Under conventional fertilizer practices, plant tissue contains high levels of nitrate which is a good indicator of the nutrient status of the plant. Soils also contain varying amounts of ammonium forms of nitrogen, which bacteria convert to nitrate forms over time. But Cardy ion meter do not measure ammonium nitrogen and therefore some underestimate of the nitrogen may become available to the plants during the growing season. The Cardy ion meter will measure the leaf nitrate-N concentration (in ppm or mg kg^{-1} fresh weight) but does not measure leaf amino acid and protein level. A portable Cardy ion meter with selective electrode has recently been developed that can directly measure

NO_3^- and NO_3^---N form of nitrogen present in fresh samples of squeezed petiole sap. This Cardy ion meter offers immediate results of in-season crop N status. Therefore, adjustments in N fertilization can be made before the crop experiences N deficiencies or excessive N applications which may lead to enhanced vegetative growth, yield reductions, and/or delayed maturity.

Fig. 1. Measurements of Chlorophyll index on potato leafs by Chlorophyll meter.

Fig. 2. Measurements of NO_3-N from petiole sap of potato crop by Cardy ion meter.

5. Materials and methods

The trial fields with potato variety planted on three different location fertilized with increasing N rate were conducted in Bosnia and Herzegovina in 2007. Data were obtained using the rapid diagnostic tools for plant material analysis in comparison to the standard measurement.

5.1 Experimental stations

Trial field were performed on three different locations Mostar (L1), Malo Polje (L2) and Stolac (L3) away from each other around 20 km. Each experimental station was set up on 300 m^{-2} of surface. All three locations are in Herzegovina region and survey was conducted in 2007. In trial field we have used common potato varieties Adora, Liseta and Romano. Potato seed were machinery sown with in row seed space 0.18-0.20m and between rows seed space 0.65m. Three fertilization treatments (0, 100, 200 kg N ha^{-1}) were used in a split-plot design with three replications. N fertilization treatments were estimated according to the soil analyses, adding the one half of the total N amount before planting and other 45 days after emergency of crop. For basic fertilization we have used NPK formulation 7:20:30 with additional dressing UREE and KAN. Potato plants grown under various N fertilizers were sampled at the 4th growth stages (65, 75, 85 and 95 days after sowing-DAS).

5.2 Collection of the samples

Chlorophyll and Cardy ion meters are used in field trial with potato crop tasting the effectives of tools in evaluation of fertilization rate. The youngest fully-expanded leaf (3-4 from the top of the canopy) was used for obtaining the values measured by Chlorophyll Meter (30 readings are collected for average values per sample). Chlorophyll measurements were preformed during the morning period when the temperatures are not high starting with sampling from the period of appearance the first flowers (65 DAS). Each SPAD value was the mean of the measurement on 6 leaflets. After this samples were stored in paper bag to the small hand fridge and later in laboratory we have proceed measurement of the NO_3-N values from the petioles sap of the same samples. From all leaflets petioles were removed by cutting and squeezing the sap by hydraulic plant press. The petiole sap was used for obtaining the values of NO_3-N concentrations by the Cardy ion meter. Both methods are compared to the standard laboratory N (percentage of total leaf N expressed on dry weight basis) measurement expecting a reliable data on N plant status as these methods are not yet tested for each crop and values can varied depending on environmental conditions. Providing the reliable data by at least one of these meters we can replace the long lasting and expensive laboratory plant sample measurement. Nitrogen concentration in plants is normally determined by expensive and time consuming chemical analyses (AOAC, 1970). As an alternative, Chlorophyll and Cardy ion meter readings in leaf and petiole sap were proposed, but these assays are not always satisfactory.

5.3 Plant material

For plant material we have used common potato varieties on Herzegovina market: cultivar Adora, Liseta and Romano. Cultivar Adora has very short vegetation about 80 days. Tubers are oval, light yellow skin colored and smooth skin with medium shallow eyes. Plants are erected medium high with smaller number of thinly shoots. Leafs are larger, dark green, half-open. This cultivar could have a high yield smaller tuber number per plant. Dry matter of tuber is lower. Cultivar Liseta has early vegetation from 95 to 100 days. Tubers are elongated-oval, smooth to medium smooth skin, light yellow skin colored and with shallow eyes. Plants are half-erected, medium-high with light colored shoots. Leafs are light green medium-open. This cultivar could have a high yield with higher tuber number per plant. Dry matter of tuber is higher. Cultivar Romano has same vegetation period as cultivar Liseta (95-100 days). Tubers are around and oval, smooth light red colored skin with relative depth eyes. Plants are half-erected with short to medium short shoots intensive colored. Leafs are larger, light green, half-open. This is a high yield cultivar.

5.4 Soil and weather analyses

Soil sampling on L1 shows neutral to medium alkali pH reaction (7.79 in H_2O and 6.59 in KCl). Level of humus at 2.55% was not satisfying. Relative low NH_4 of 0.98 mg per 100 g soil and NO_3 of 0.32 mg per 100 g soil have measured. Level of P_2O_5 of 35.9 mg per 100 g soil and K_2O of 24.5 mg per 100 g soil were satisfying. Anthropogenic soil on alluvial deposit with sandy loam texture was noticed.

Soil sampling on L2 shows neutral to medium alkali pH reaction (7.67 in H_2O and 6.43 in KCl). Level of humus at 2.55% was low. Relative low NH_4 of 0.98 mg per 100 g soil and NO_3 of 0.32 mg per 100 g soil have measured. Level of P_2O_5 of 17.5 mg per 100 g soil and K_2O of 20.0 mg per 100 g soil were satisfying. Soil texture on this location is classified as clay loam.

L3 shows medium acid pH soil reaction (6.56 in H_2O and 5.11 in KCl). Level of humus at 2.41% was low. Relative low NH_4 of 0.78 mg per 100 g soil and NO_3 of 1.24 mg per 100 g soil have measured. Very poor level of P_2O_5 of 0.40 mg per 100 g soil was detected while K_2O of 18.4 mg per 100 g soil were satisfying. Heavy mechanical soil composition on this location was noticed with poor water/air ratio and it was classified as loamy to loamy clay soil type.

Weather condition on three locations was measured from nearest local weather station. Colleted data on average monthly temperatures (°C), precipitation (L m-2) and air humidity (%) were shown for location Mostar (Figure 3), Malo Polje (Figure 4) and Stolac (Figure 5) in text.

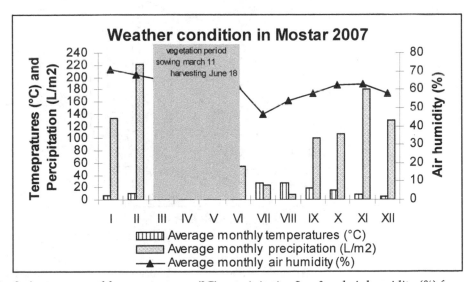

Fig. 3. Average monthly temperatures (°C), precipitation L m-2 and air humidity (%) for Mostar with marked gray part for vegetation period form sowing date March 11 till harvesting June 18

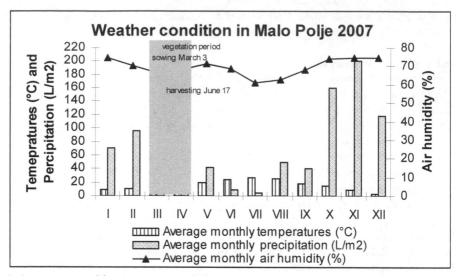

Fig. 4. Average monthly temperatures (°C), precipitation L m-2 and air humidity (%) for Malo Polje with marked gray part for vegetation period form sowing date March 3 till harvesting June 17

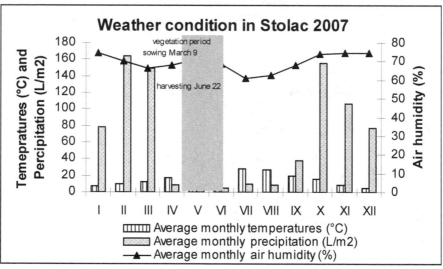

Fig. 5. Average monthly temperatures (°C), precipitation L m-2 and air humidity (%) for Stolac with marked gray part for vegetation period form sowing date March 9 till harvesting June 22

For each location Figures were shown also with data about vegetation periods with information on sowing and harvesting date. For L1, L2 and L3 sowing dates were on March 11, 3 and 9 and harvesting were on June 18, 17 and 22. During season 2007 dry weather period has occurred in phase on tuber initiation. The phase of tuberization and tuber soaking were followed by increasing precipitations period with final phase of tuber maturation amount of precipitation was decreased. Small increment in all three measured weather parameters were noticed in L3 for main growing stages while L1 and L3 shows relative similar situation. Comparison of obtained values with ten years average data shows slight increment in all three measured parameters.

5.5 Statistical analyses

Analyses of variance (ANOVA) were used for testing differences in SPAD values, concentration of NO_3-N mg kg^{-1} and percentage of total nitrogen in potato leaves. Pearson's correlation coefficient (*, **, p <0.05 and p <0.01) model was used for identifying correlations between: the total nitrogen (determinate on dry matter basis) and values collected with meters. Data were analyzed by using SPSS for Windows v 13.0 (SPSS, 2004).

Our research objective was to determine how accurately the two meters estimated potato leaf N concentration. Chlorophyll (SPAD-502 meter, Minolta, Osaka,. Japan) and Cardy ion meter readings were compared with actual leaf N concentration (obtained using the Kjeldahl method) determined by laboratory plant analysis.

6. Results and discussion of plant analysis measurement on potato crop

Chlorophyll and Cardy ion readings have follow the nitrogen fertilization rate showing the good correlation relations to the total nitrogen measured in same samples (Table 1). Changes in values were expected regarding a variety and experiment location. Irrigation of the trial field was also provided according to the weather conditions.

Correlation coefficient (r) between values	Days after sowing (DAS)				
	65	75	85	95	Locations
N & NO_3-N	0.88**	0.71**	0.41*	0.24ns	Jasenica
	0.59**	0.59**	0.73**	0.14ns	Malo Polje
	0.62**	0.91**	0.85**	0.53**	Solac
N & SPAD	0.73**	0.59**	0.45ᴬ	0.43*	Jasenica
	0.37ns	0.67**	0.75*	0.32ns	Malo Polje
	0.31ns	0.75**	0.65*	0.73**	Solac
SPAD & NO_3-N	0.91**	0.75**	0.73**	0.46*	Jasenica
	0.61**	0.59**	0.64**	0.54**	Malo Polje
	0.35ns	0.72**	0.71**	0.68**	Stolac

Table 1. Correlations (r) between total N analysed on dry mater basis (Kjeldahl method) of potato leaf and NO_3-N values detected by Cardy-ion meter in petiole sap of potato; Correlations (r) between total N analysed on dry mater basis of potato leaf and SPAD values detected by chlorophyll meter; Correlation (r) between NO_3-N and SPAD values on three different locations during four sampling period (ns-non significant; * - significant at P=0.05; significant at the P=0.01)

Correlation relationships between SPAD and NO_3-N values obtain at all three location shows very significant coefficient for every sampling period. Other comparisons between N and NO_3-N as well and N and SPAD values haven't show positive coefficient during the all sampling period. Usually non significant coefficients were spotted at the beginning or at the end of the sampling period. The amount of chlorophyll a and b has been investigated in many studies and shown to correlate closely with leaf %N and SPAD meter value (Neukirchen et al., 2002). Vos & Bom (1993) have carried out the experiment with potato varieties Vebeca fertilized with 0, 110, 180 and 250 kg N ha^{-1} in split application and they have compared a data of SPAD and NO_3-N values with standards. From the results they have confirmed a good correlation relation between SPAD values and chlorophyll with coefficient r=0.97. They have also confirmed good correlations between SPAD values and N while between SPAD and NO_3-N values coefficient was low. Gianquinto et al. (2004), in their investigation center in Scotland try to find a strong link between SPAD values and total leaf nitrogen. Very high correlation coefficient was established during the middle of the vegetation season while at the beginning and end of the season relationship was week.

This survey has try to identify the nitrogen concentration using a rapid diagnostic tools for plant analysis as SPAD meter and nitrate level by Cardy ion meter in potato leaf during a different growth stages. Results of the ANOVA test for nitrogen concentration in potato leaf are shown in table 2.

	Days after sowing (DAS)			
Source of variability	65	75	85	95
	F-Test			
Location (L)	**	*	**	**
Cultivar (C)	**	**	**	*
Fertilization (F)	**	**	**	**
L × C	*	ns	ns	ns
L × F	ns	*	**	**
C × F	ns	ns	ns	ns
L × C × F	ns	ns	ns	ns
Fertilization treatment				
Control	3.53	3.19	2.86	2.38
100 kg N ha^{-1}	4.18	3.54	3.09	2.52
200 kg N ha^{-1}	4.63	4.04	3.44	2.87
$LSD_{0.05}$	0.20	0.14	0.10	0.15
Cultivars				
Adora	3.90	3.42	3.01	2.37
Liseta	4.22	3.70	3.24	2.66
Romano	4.21	3.65	3.14	2.74
$LSD_{0.05}$	0.13	0.16	0.05	0.27

Table 2. Result of ANOVA test for N concentration in potato leaf during the 4 growth stages. (ns - non significant; * - significant at P=0.05; significant at the P=0.01)

Beside significant impact on N leaf concentration of all three factor used in the experiment (location, cultivar and different fertilization rate) it is very interesting to note interaction of location and cultivar for 65 DAS as well as interaction of location and fertilization treatment for other three growing periods (75, 85 and 95 DAS). This means for 65 DAS that the values of the nitrogen at all three locations and for all three cultivars has varied. It was also conclude that the interaction for 75, 85 and 95 DAS shows significant different N concentration in potato leaf for all three location and the nitrogen fertilization treatments have achieved different values in each location. Results of the ANOVA test for SPAD values in potato leafs are also shown in the table 3.

As for nitrogen it is noted significant impacts of each factor on SPAD values. Interaction of location and cultivar was achieved at the beginning and end of the season while interaction of location and fertilization treatment was significant for each growth stage. The first interaction means that the each cultivar shows different SPAD values for each location. This means that the each location with them specific microclimate and soil conditions could affect the results. The second interaction indicates that different fertilization treatments at each location caused the differences in SPAD values. Generally the L3 has obtained highest SPAD values, followed by L1 while L2 was usually presented for each growing stage with smallest SPAD values.

	Days after sowing (DAS)			
Source of variability	65	75	85	95
	F-Test			
Location (L)	**	**	ns	**
Cultivar (C)	**	**	**	**
Fertilization (F)	**	**	**	**
L × C	**	ns	ns	*
L × F	**	*	**	**
C × F	ns	ns	ns	ns
L × C × F	ns	ns	ns	ns
Fertilization treatment				
Control	44.6	38.9	37.1	32.4
100 kg N ha^{-1}	48.1	43.3	40.8	35.1
200 kg N ha^{-1}	49.7	46.2	44.7	38.3
LSD$_{0.05}$	0.64	0.99	1.31	1.35
Cultivars				
Adora	48.8	41.8	40.5	35.0
Liseta	45.2	40.8	39.2	33.6
Romano	48.4	45.8	42.8	37.2
LSD$_{0.05}$	0.68	1.25	1.17	0.96

Table 3. Result of ANOVA test for SPAD values in potato leaf during the 4 growth stages at all three locations. (ns - non significant; * - significant at P=0.05; significant at the P=0.01)

As we have obtained results of ANOVA test for N concentration and SPAD values, we have proceeded with the same test for NO_3-N values. Results of the ANOVA test for NO_3-N values measured by Cardy ion meter in potato leaf are also shown in the table 4.

NO_3-N values are under the strong interaction between location and fertilization treatments while impact of location and cultivar interaction is noticed only on sampling at 85 DAS. Single factor impact on NO_3-N values for each growing stages was more expressed for location and for fertilization treatment while cultivar was significant only at the beginning of the season.

Source of variability	Days after sowing (DAS)			
	65	75	85	95
	F-Test			
Location (L)	**	*	**	**
Cultivar (C)	**	ns	ns	ns
Fertilization (F)	**	**	**	**
L × C	Ns	ns	*	ns
L × F	**	**	**	**
C × F	Ns	ns	ns	ns
L × C × F	Ns	ns	ns	ns
Fertilization treatment				
Control	543	244	107	154
100 kg N ha-1	1469	672	454	375
200 kg N ha-1	2069	1714	1186	985
$LSD_{0.05}$	160,4	201,4	144,6	229,9
Cultivars				
Adora	1481	860	614	683
Liseta	1418	852	576	352
Romano	1206	919	557	479
$LSD_{0.05}$	108,2	-	-	-

Table 4. Result of ANOVA test for NO_3-N level in petiole sap during the 4 growth stages at all Three locations. (ns - non significant; * - significant at P=0.05; significant at the P=0.01)

Average values shown in the table for N, SPAD and NO_3-N values were highest at L3. L1 has lower average values of N for 10%, SPAD for 9% and NO_3-N 54%. Comparing to L3 lower average values also has L2 for N 16%, SPAD 15% and NO_3-N 52%. If these measurements can serve to evaluate crop nutrient status it is logical that increscent or decrement of these values can influenced on crop yield. Even the average yield for each location statistically is not differing; the highest yield has achieved at L1, opposite from the expected highest results of measured values on location L3. These facts are explaining with environmental conditions on each location which were affected the nutrient availability and genetic potential of the crop. Cultivar Adora has average tuber yield of 23.30 t ha-1, cultivar

Romano 25.57 t ha⁻¹ and Liseta has obtained the highest yield of 26.86 t ha⁻¹. Each location with their specific microclimate conditions can affect the yield formation of each cultivar which was confirmed interaction occurred between location and cultivar (graph 1). Same nutrient fertilization can result in different yield formation regarding to the location soil characteristic or weather conditions which was also confirmed interaction between location and fertilization treatments (graph 2). Also, genetic potential of each cultivar (vegetation period) can be affected by conditions on each location (e.g. temperature, light, precipitation, air moisture ect. (Gianquinto et al., 2006). Besides length of the potato growing period, early and late potato varieties differ in their dry matter accumulation and N assimilation rate (Kleinkopf et al., 1981). Reason for higher measured values at L3 can be explained by different soil reaction where this location has lower pH reaction comparing to other two. L3 have low acid to neutral soil pH reaction which was better for potatoes growing while other two locations have higher pH values. Average values of N, SPAD and NO₃-N were highest for first two sampling period and after that it was noticed slight decrement in values as the vegetation season passing. Reason for values diminishing with the time was that at beginning of season plants have intensive photosynthesis afterward they redistribute the nutrient from the canopy in the storage parts (tuber) which decrease protein content in leafs and increase in potato tuber (Millard & MacKerron, 1986). From our survey we can noticed that the cultivar with shortest vegetation period Adora has lowest N, SPAD and NO₃-N values because its earlier starts with redistribution of nutrient from canopy to the underground plants parts. Generally, earliest cultivar might required a different fertilization managements comparing to the latest potato varieties since tuberziation and N uptake occurs earlier in the season if they planted at the same time. In this case the amount of available N in soil can be very low when tuberization in later cultivar has occurred. A high application of N rate at the planting period should be avoided since the high amount of available nitrogen can delay potato tuber formation for 7 to 10 days (Kleinkopf et al., 1981) promoting the vegetation growth. This would be particularly important in areas with limited growing season.

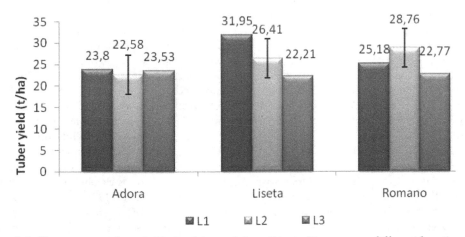

Graph 1. The average tuber yield of cultivars Adora, Liseta, Romano on different locations L1, L2 and L3

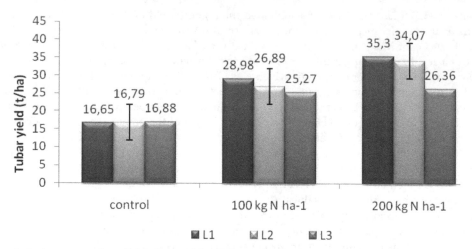

Graph 2. Average tuber yield obtained by different fertilization rate on three location L1, L2 and L3

Gianquinto et al., (2004) have found that early cultivar shows higher SPAD values comparing to the later cultivar even the measurements were provided in the same period. From the data presented here, SPAD measurements were varied depending on cultivar. We have Romano cultivar with highest SPAD values, lower values have Adora and lowest were spotted in cultivar Liseta. Even Liseta cultivar showed highest N content the SPAD values were lowest while other two cultivars haven't shown significant variations in SPAD values. Reason for this was measurement procedure which implies recording a SPAD values in first full develop but not total matured leafs. Not all cultivar have same capacity for nutrient accumulation. If the measurement were not provided in the same vegetation period certain differences between total leaf N and SPAD values can be recorded even we have same samples. Debaeke et al. (2006) have conclude that the SPAD values highly depend on cultivar type, because not all cultivar have same thickens of the leaf as the thicker leaf shows higher chlorophyll units per leaf but not necessarily and higher N values. Beside cultivar type high impact on SPAD values have soil type, climate, water status in soil and plant. According to the Wheeler et al. (1989) decrement of the moisture content increase leaf surface and intensify leaf colour causing a higher SPAD values. In this survey applied fertilization treatments from 0 to 200 kg N ha-1 showed average SPAD values for 65 DAS from 44,5 to 49,5; 75 DAS from 38,8 to 46,2; 85 DAS from 37,1 to 44,7; 95 DAS from 32,4 to 38,3 unit. Increasing nitrogen fertilization rate at all three locations have obtained increment of SPAD values and N values in crop also the values have diminished white time. These results were interpreted as good response of SPAD meter on N fertilization treatments and meter can serve as good tool in the evaluation of the plant nutrient status especially if we use a small control plot as reference strip. Nutrient concentrations decline ontogenetically during the growth period, even with sufficient N supply (Siman 1974; Sorensen, 2000). Therefore, the critical concentration has to be related to a carefully defined growth stage (Lorenz & Tyler, 1977). However, the way the growth stages are defined is often imprecise. Differences between SPAD and N values obtained in potato leaf samples could be explained by weather condition. In year when the experiment was set up during the vegetation period

(April, May and June) the temperatures were very high with low precipitation. In period of tuber initiation the temperatures have exceed a ten years average (13.3°C in April) for 4°C. According to the Wheeler et al., (1989) decreased moisture content could cause increment of leaf surface and intensify leaf colour. As we have mentioned the sensitivity of SPAD meter on environmental conditions the high temperatures with low precipitations could cause small transpiration rate of nutrient and water trough the plant resulting in the small intensity of photosynthesis. Stress condition for plant can develop increment of SPAD values at same level as plants that have received a fertilization treatments and water supply (Gianquinto et al., 2004).

The Cardy ion measurement can also provide a currently nutrient status in plants, detecting the nitrate-nitrogen content that was not yet incorporated in to organic compounds. In potatoes the Cardy ion measurements were widely accepted procedure to enable a quick assessment of the crop canopy nitrogen N status to derive N-fertiliser recommendations. Generally, Cardy ion measurements are more sensitive for detecting the N status of a crop compared to the chlorophyll analysis. This is because the nitrate concentration reacts more rapidly than the total N concentration to changes in the N supply (Huett and White, 1992). Cultivar with shorter vegetation period Adora and Liseta have shown the higher NO_3-N content only for measurement on 65 DAS while after Romano (longer vegetation period) variety have Obtained higher values as the varieties were not in the same vegetation phase. As the nitrate concentration varied regarding to the type of cultivar we have also found a good response of NO_3-N values on N fertilization treatment. Each treatment has shown differences in the NO_3-N values. Comparing to SPAD meter where values were smaller so the differences were mainly recorded between the fertilized treatment and control (no fertilization). The NO_3-N values for 65 DAS were from 543 to 2069 mg kg^{-1} fresh weight; for 75 DAS from 244 to 1714 mg kg^{-1}; for 85 DAS 107 to 1186 and for last period of sampling 95 DAS we reached the values from 154 to 985 mg kg^{-1} fresh weight. According to the Love et al., (1999) fertilization treatments from 0 to 250 kg N ha^{-1} applied to Russet Burbank potato variety have obtained NO_3-N concentration from 22000 to 24000 mg kg^{-1} dry weight in period of early tuberization and at the end of the vegetation season the values were from 2000 to 6000 mg kg dry weight. Differences between measurement in fresh and dry matter are large. For the Russet Burbank NO_3-N concentration in fresh weight have flouted from 1300 to 1600 mg kg^{-1} in early maturation period while at the end of the vegetation the concentration have decreased on 550 to 700 mg kg^{-1} fresh weight. In our experiment small variation were noticed for the NO_3-N values recorded in petiole sap during the four growth stages. On 85 DAS values shows lower NO_3-N concentration then on the last sampling period 95 DAS which was not in the accordance to the pervious statement of diminishing values as vegetation period ends. We explain this with strong influence of environmental condition where at the end of the season the higher precipitations have occurred. Under normal growing condition (with no excessive or any precipitation, extremely high or low temperatures and irradiation) nitrate nitrogen taken up by plants is readily converted into amino acids and proteins. Hence, the level of nitrate is not high enough to be toxic. During prolonged periods of moisture stress (drought) and high temperatures or low humidity, nitrate accumulates in plants (Malakouti, 2002). The severity of nitrate accumulation in plants is accentuated by heavy nitrogen fertilization prior to the onset of drought. Soil N mineralization, N fertilization rate, sampling date, and cultivar can influence the petiole N level (Vitosh & Silva, 1996). Insufficient water for plant growth can result in the accumulation of NO_3-N in petioles (Meyer & Marcum, 1998). Potatoes that have not received current season nitrogen fertilization generally have lower petiole N values earlier in the season and these values decline rapidly after tuber set (Gardner & Jones, 1975; Porter & Sisson, 1991; Wescott et al., 1991). It has been suggested since the rate of petiole N change during the season so rapid establishing of critical petiole N would be difficult unless the precise age of the plant is known and since the soil NO_3-N status less fluctuates it might could be the better indicator of the N status of the crop (Doll et al., 1971; Rodrigues, 2004). More often sampling in closer intervals earlier in the season might have facilitated interpretation of petiole N trends.

According to the location influence on nitrate values the highest were recorded at L3 followed by L1 and the lowest were at L2. Fertilization treatments have significantly influenced on NO_3-N values but the interaction of fertilization and location has shown that same fertilization treatments have achieved the highest values at L3. The lowest values of all measured parameters on L2 are defined by vary unfavorable microclimatic conditions where no participations have occurred followed by high temperatures during intensive potato growth. This variation in values between locations were rather connect to the weather then soil conditions where soil analyses have shown similar nutrient and other soil parameters at L1 and L2. So according to the soil parameters differences in values should not be occurred. Regarding to the soil nutrient availability it is important beside N application rate the residue cover, soil aggregation and structure, soil organic matter, crop rotation. These are the factor which can often result in lower nitrogen application rate, better use of applied nitrogen or less nitrate leaching.

A monitoring program is suggested due the variability in values from one site to the next and in order to measure the actual effects of soil N availability during growth (Wescott et al., 1991). Petioles have often been used for quick tests to estimate nitrate-N in sap (tissue nitrate-N). The quick tests have been developed for field use to avoid the time lag between sampling and result as well as the costs of laboratory analysis. Petiole plus midrib nitrate has been shown to reflect the N status, as for example in potato (Bélanger et al., 2003). Generally, nitrate concentrations tend to be higher in stems and petioles than in leaves so the nitrate level of vegetables that their petioles, stems or leaves are consume must be considered. Nitrate accumulation in older organs is high because activity of nitrate reductase in these parts is low (Malakouti, 2002). Nitrate and nitrite are detrimental components in plants after applying of nitrogen fertilizers in soil and oxidation by microorganism, the produced nitrate have high affinity to absorption by plants (Hogg et al., 1992; McKnight et al., 1997).

7. Conclusion

Data presented here shows response of the potato plant to the available nitrogen rate as an important factor for accurate fertilization practice. Obtained data confirm and extended previous studies which showed that chlorophyll measurement in combination with Cardy ion meter can be used in field conditions to detect differences in the response of a potatoes genotype to nitrogen supply. However, growers should understand that the chlorophyll meter is only able to reveal deficiency situations. Cardy ion meter readings are accurate enough to be used on a practical basis as a decision-making tool that can increase the efficiency of fertilizer use. If these two meters are calibrated correctly, it's possible to conclude plant required N amount. Chlorophyll meter and Cardy ion meter are most effective when used in conjunction with other agronomic tools or agronomic knowledge. We also suggest that the N fertilization recommendations should be develop for each potato varieties. This practice for chlorophyll and Cardy ion measurement should be useful in management strategies which maximize use of previous crop residues, organic amendments and soil reserves as N sources. In such a system, at-planting N fertilizer applications would be reduced and supplemental N application could be applied when values of leaf chlorophyll or nitrate monitoring are beyond critical one for analysed crops taking care about environmental impact to recorded values. Analytical procedures for soil and plant analyses widely varied from location to location including and other factors (weather conditions, crop variety etc.); it is important to use test procedures calibrated for each geographical area or crops of interest. These nitrogen-monitoring techniques enable growers to apply nitrogen fertilizer at the right time, helping to ensure high yields without making unnecessary applications that could adversely affect the environment or possibly increase some pest damage. This also can save farmers money. The principal advantage to these techniques is the ability to get immediate results. Other advantages are that the equipment is portable and easily available through catalogs. There are several disadvantages to using the meter. First, the results, though close, are not as accurate as measurements from an analytical laboratory. Second, the instruments are sensitive to environmental conditions and can give inaccurate or inconsistent readings if exposed to heat, direct sunlight, water, sand and dirt. Therefore testing of sap should take place indoors under controlled environmental conditions and the reader should be kept clean and dry while chlorophyll reading should be immediately preformed. The environmental conditions also include those that directly affect the plant, which include excessive rain or irrigation at time of testing. Additional research is needed to determine the optimum amount of N that can be applied for certain potato variety.

8. References

Association of Official Analytical Chemists. (1995). Official methods of analysis of the Association of Official Analytical Chemists. AOAC, Washington, DC. C. 34, p. 8

Bélanger, G.; Ziadi, N.; Walsh, J.R.; Richards J.E.; Milburn P.H. (2003). Residual soil nitrate after potato harvest. J. Environ. Qual. 32:607–612.

Bishop, R.F. & MacEachern, C.R. (1971). Response of spring wheat and barley to nitrogen, phosphorous and potassium. Can. J. Soil Sci. 51:1–11.

Carter, J.N. & Traveller D.J. (1981). Effect of time and amount of nitrogen uptake on sugar beet growth and yield. Agron. J. 73:665–671.

Cole, D.F.; Halvorson, A.D; Hartman, G.P.; Etchevers, J.E.; Morgan, J.T. (1976). Effect of nitrogen and phosphorous on percentage of crown tissue and quality of sugar beets. N.D. Farm Res. 33(5) :26 –28.

Davidson, E. A.; (2009).The contribution of manure and fertilizer nitrogen to atmospheric nitrous oxide since 1860, Nat. Geosci., 2, 659–662.

Debaeke, P.; Rouet, P.; Justes, E. (2006). Relationship between the Normalized SPAD Index and the Nitrogen Nutrition Index: Application to Durum Wheat. Journal of Plant Nutrition, 29: 75–92.

Directive 98/83/EC (1998) on the quality of water intended for human consumption. (Drinking Water Directive).

Doll, E.C., Christensen, D.R. & Wolcott A.R. (1971). Potato yields as related to nitrate levels in petioles and soils. Am. Potato J. 48:105-112.

Dorn, T; (2011). Soil Fertility-Nitrogen. Natural sources of nitrogen for plant growth. The Farm view. The nebline. Pp2. http://lancaster.unl.edu

Ekbladh, G. (2007). Plant Analysis as a Tool to Determine Crop Nitrogen Status; Towards Leaf Area Based Measurements. Doctoral thesis, Swedish University of Agricultural Sciences, Uppsala

Erisman, J.W.; Sutton, M.A.; Galloway, J; Klimont Z. & Winiwarter, W. (2008). How a century of ammonia synthesis changed the world. Vol. 1 (Oct.): 636-639.

Errebhi, M.; Rosen, C.J.; Gupta, S.C.; Birong, D.E. (1998). Potato yield response and nitrate leaching as influenced by nitrogen management. Agron. J.90 :10 –15.

European Environment Agency (2001). Eutrophication in Europe's coastal waters, EEA, Copenhagen.

European Environment Agency (2005). The European environment: state and outlook 2005, EEA, Copenhagen.

Fageria, N.K. & Baligar, V.C. (2005). Enhancing Nitrogen Use Efficiency in Crop Plants. Advences in Agronomy, Vol 88, pp 97-185.

FAO (2010). United Nations Food and Agricultural Organization. FAOSTAT database http://faostat.fao.org

Gardner, B.R. & Jones, J.P. (1975). Petiole analysis and the nitrogen fertilization of russet burbank potatoes. Am. Potato J. 52:195-200.

Gianquinto, G.; Goffart, J.P.; Olivier, M.; Guarda, G.; Colauzzi, M.; Dalla Costa, L.; Delle Vedove, G.; Vos, J.; MacKerron, D.K.L. (2004). The use of hand-held chlorophyll meters as a tool to assess the nitrogen status and to guide nitrogen fertilization of potato crop. Potato Res. 47:35–80.

Gianquinto, G.; Sambo, P.; Borsato, D. (2006) Determination of SPAD threshold values in order to optimise the nitrogen supply in processing tomato. Acta Hort.700, 159-166.

Grizzetti, B. , Bouraoui, F. and De Marsily , G. (2008). Assessing nitrogen pressures on European surface water. Global Biogeochemical Cycles , 22 , GB4023

Hill, W.A. (1984). Effect of nitrogen nutrition on quality o f three important root / tuber crops. Pp.627–641. In Nitrogen in crop production. R.D. Hauck (ed.). ASA/CSSA/SSSA Madison, WI.

Hills, F.T. & Ulrich, A. (1971). Nitrogen nutrition. Pp. 111–115. In R.T. Johnson et al., (ed.) Advances in sugarbeet production: principals and practices. Iowa State Univ. Press, Ames, Iowa.

Hogg, N.; Darley-Usmar, V.M.; Wilson, M.T. ; Moncada, S. (1992). Production of hydroxyl radicals from the simultaneous generation of superoxide and nitric oxide. Biochem. J., 281: 419-424.

Huett, D. O., White, E. (1992). Determination of critical nitrogen concentrations of potato (Solanum tuberosum L. cv. Sebago) grown in sand culture. Aust. J. Exp. Agric. 32, 765–772.

Jones, Jr.J.B.; Wolf, B.; Mills, H.A. (1991). Plant Analysis Handbook. Micro Macro Pub. Athens. pp: 39-43,99-104, 178-187.

Kelling, K.A.; Combs, S.M.; Peters, J.B. (2000). Plant Analysis as a diagnostic tool. http://www.soils.wisc.edu/extension/publications/ horizons/2000/Plant%20Analysis%20as%20Tool.pdf.

Kleinkopf, GE.; Westermann, DT.; Dwelle, RB. (1981). Dry matter production and nitrogen utilization by six potato cultivars. Agron J 73:799-802.

Laughlin, W.M. (1971). Production and chemical composition of potatoes related to placement and rate of nitrogen. Am. Potato J. 48:1-15.

Lewis, D.C.; Grant, I.L.; Maier, N.A. (1993). Factors affecting the interpretation and adoption of plant analysis services. Australian Journal of Experimental Agriculture 33, 1053-1066.

Locascio, S.J.; Wiltbank, W.J.; Gull, D.D.; Maynard D.N. (1984). Fruit and vegetable quality as affected by nitrogen nutrition. Pp. 617-626. In Nitrogen in crop production. R.D. Hauck (ed.). ASA/CSSA/SSSA Madison, WI.

Lorenz, O.A. & Tyler, K.B. (1977). Plant tissue analysis of vegetable crops. Division of Agricultural Sciences, University of California. Bulletin No 1879, 21-24.

Love, S.L.; Bohl, W.H.; Corsini, D.; Stark, Jeffery C.; Olsen, N.; Pavek, J.; Mosley, A. (1999). Cultural Management of Bannock Russet Potatoes.

Macy, P. (1936). The quantitative mineral nutrient requirements of plants. Plant Physiol., 2: 749-64

Malakouti, MJ., (2002). Evaluation of N-fertilizers effects on nitrate accumulation in vegetables. Final report. Agricultural Research and Education Organization. Agricultural Commission, National Council for National Scientific Research. Tehran, Iran.

McKnight, G.; Smith, L.M.; Drummond, R.S.; Duncan, C.W.; Golden M.N.H.; Benjamin, N. (1997). The chemical synthesis of nitric oxide in the stomach from dietary nitrate in man. Gut, 40: 211-214.

Meyer, R.D. & Marcum, D.B. (1998). Potato yield, petiole nitrogen, and soil nitrogen response to water and nitrogen. Agron J 90:420-429.

Millard, P. & MacKerron, D.K.L. (1986). The effect of nitrogen application on growth and nitrogen distribution within the potato canopy. Annales of Applied Biology, 109, 427-37.

Neukirchen, D. & Lammel, J. (2002). The chlorophyll content as an indicator for nutrient and quality management. Nawozy i Nawozenie – Fertilisers and Fertilisation 2, 89-109.

Olesen, J.E.; Carter, T.R.; Diaz-Ambrona, C.H.; Fronzek, S.; Heidmann, T.; Hickler, T.; Holt, T.; Minguez, M.I.; Morales, P.; Palutikov, J.; Quemada, M.; Ruiz-Ramos, M.; Rubæk, G.; Sau, F.; Smith, B.; Sykes, M. (2007). Uncertainties in projected impacts of climate change on European agriculture and ecosystems based on scenarios from regional climate models. Clim. Change 81, 123-143.

O'Leary, M.; Rehm, G. Schmitt, M. (2002). Understanding Nitrogen in Soils. University of Minnesota. U.S. Department of Agriculture, Extension Service, under special project number 89-EWQI-1-9180

Painter, C.G.; Ohms, R.E.; Walz, A. (1977). The effect of planting date, seed spacing, nitrogen rate and harvest date on yield and quality of potatoes in Southwestern Idaho. Univ. of Idaho Agric. Exp. Stn. Bull. No. 571.

Peng, S. (1993). Adjustment for specific leaf weight improves chlorophyll meter's estimate of rice leaf nitrogen concentration. Agron. J. 85:987–990.

Porter, G.A. & Sisson, J.A. (1991). Petiole nitrate content of Maine-grown Russet Burbank and Shepody potatoes in response to varying nitrogen rate. Am. Potato J. 68:493-505.

Rashid, A. (2005). Soils: Basic concepts and principles. In: Soil Science. Memon, K.S. and A. Rashid, (eds.). National Book Foundation, Islamabad.

Rodrigues, M.A. (2004). Establishment of continuous critical levels for indices of plant and presidedress soil nitrogen status in the potato crop. Communications in Soil Science & Plant Analysis 35:2067-2085.

Röling, N. & Wagemakers A. (2000) Facilitating Sustainable Agriculture: Participatory learning and adaptive management in times of environmental uncertainty. Cambridge University Press, Cambridge, 318p

Schröder, J.J.; Neeteson, J.J.; Oenema, O.; Struik, P.C. (2000). Does the crop or the soil indicate how to save nitrogen in maize production? Reviewing the state of the art. Field Crops Research 66, 151-164.

Siman, G., (1974). Nitrogen status in growing cereals, with special attention to the use of plant analysis as a guide to supplemental fertilization. Doctoral thesis. The Royal Agricultural College of Sweden, Uppsala. 93 pp.

Smeal, D. & Zhang, H., (1994). Chlorophyll meter evaluation for nitrogen management in maize. Soil Science and Plant Analysis, 25: 1495-1503.

Sorensen, J.N. (2000). Ontogenetic changes in macro nutrient composition of leaf-vegetable crops in relation to plant nitrogen status: A review. Journal of Vegetable Crop Production 6, 75-96.

Spectrum technologies, Articles (2011) Walter Ridell -Instrument for Rapid Nitrogen Testing in Corn Nitrate Meter and Chlorophyll Meter.

(http://www.specmeters.com/pdf/articles/nitrogen_testing_corn.pdf)

Vitosh, M.L. & Silva, G.H. (1996). Factors affecting potato petiole sap nitrate tests. Communications in Soil Science & Plant Analysis 27:1137-1152.

Vos, J. & Bom, M. (1993). Hand-held chlorophyll meter: A promising tool to assess the nitrogen status of potato foliage. Potato Res. 36: 301–308.

Ward, M.; de Kok, T.; Levallois, P. (2005). Drinking water nitrate and health: recent fi ndings and research needs. Environmental Health Perspectives 113 , 1607 –1614.

Wescott, M.P.; Stewart, V.R. ; Lund, R.E. (1991). Critical petiole nitrate levels in potato. Agron J 83:844-850.

Westermann, D.T. & G. E. Kleinkopf. (1985). Nitrogen requirements of potatoes. Agronomy Journal. July-August. 77:616 –621.

Wheeler, R.M.; Tibbitts, T.W.; Fitzpatrick, A.H. (1989). The potato growth in response to relative humidity. Hort Scienc. 24 (3) 482-484.

WHO - The world health report 2007 - A safer future: global public health security in the 21st century

Zubriski, J.C.; Vasey, E.H.; Norum, E.B. (1970). Influence of nitrogen and potassium fertilizers and dates of seeding on yield and quality of malting barley. Agron. J. 62:216 –219.

10

Marginal Waters for Agriculture – Characteristics and Suitability Analysis of Treated Paper Mill Effluent

P. Nila Rekha and N.K. Ambujam*

Soil and water conservation engineering,
Central Institute of Brackishwater Aquaculture,(ICAR), Chennai
Center for water resources, Anna University, Chennai
India

1. Introduction

With increasing global population, the gap between supply and demand for water is widening and is reaching such alarming levels that in some parts of the world it is posing a threat to human existence. Scientists around the globe are working on new ways of conserving water. On the other hand, disposal of municipal wastewater and industrial effluents are causing major environmental problem and attempts are being made all round the world to recycle and reuse it effectively and efficiently. Utilization of the marginal quality water for agricultural is in fact an appropriate technology. Agricultural scientists throughout the world are looking into the possibility of using saline and marginal quality of water for irrigation (Ayars et al., 1993; Tanji, 1997; Bajwa, 1997;Alazaba,1998;Franco et al.,2000 and Sivanappan,2000), social forestry, development of pasteur land, artificial recharge, aquaculture and wet land development. Wastewater reuse, reclamation and recycling are essential in coming years for the development of sound water and environment management policies. In arid and semi arid regions marginal water utilization is a vital component of their water resources development ensuring alternative water resources, sustainability, reduction of environmental pollution and health protection. Wastewater from different industries, which falls under marginal water quality, can be utilized beneficially for irrigation if proper treatment, monitoring and management measures were taken. The challenges and the benefits of marginal water quality utilization has to be ascertained and appropriate package of practices which are location specific needs to be followed for the real success and long term sustainability.

1.1 Wastewater irrigation - Domestic

Wastewater reuse is as old as civilization. Wastewater reclamation and reuse may produce reliable source of water even in drought years. It provides a unique and viable opportunity to augment traditional water supplies (Asano, 2002). It can help to close the loop between water supply and wastewater disposal. Moreover, nutrients beneficial to plant growth are

---

* Corresponding Author

available in the domestic waste. Irrigation water and plant nutrients being limiting factors for agricultural production in the country, exploitation of nutritional/manurial and irrigational sources from wastewater is an appropriate strategy. Many farmers use treated or untreated wastewater for irrigation. In Israel 65% of domestic wastewater is treated and used for irrigation. In fact, the 30% of irrigation is done by the wastewater and the potential is likely to be increased to 80% by 2025.(The arid and semi arid areas of the world can easily augment 15 to 20% of their water supply through reuse of wastewater (Abul, 1989). Water recycling and reuse is expanding rapidly throughout the world. Rough estimates indicate that at least 20 million hectares in 50 countries are irrigated with wastewater (Hussain et al., 2001). There are more than 1000 reuse systems in United States (Arber, 2000). Shelef and Azov (1996), illustrated the current experiences in various countries in Mediterranean region. Videla et al., (1997) has explained the experiences of wastewater treatment in Chilean forest industry. Tsagarakis et al (2001) highlighted the problems and management of wastewater in Greece. Barbagallo et al (2001) opined that the planned exploitation of municipal and industrial wastewater would help to meet the irrigation water demand in Italy. Lazarova et al (2000) highlighted the role of wastewater reuse in the development of new integrated resource management strategy in Europe and Middle East. Thus the necessity of wastewater irrigation has been realized and recognized worldwide resulting in increased expansion of wastewater irrigation programmes throughout the world.

1.2 Wastewater irrigation - Industrial

Though the concept of reusing and recycling for irrigation was an age-old practice, the industrial effluent irrigation is only of recent. It has been adopted with great vigour by most industries because the effluent standards to be met for disposal are higher when compared to land application and the pollution control measures are implemented by effective legislation and by the State Pollution Control Board. Wastewater from different industries, which falls under marginal water quality, can be utilized beneficially for irrigation with proper management. The effluents from agro based industries, which use agriculture products as the raw material, and industries, which involved in processing of agricultural products are not detrimental to the environment since the wastes are organic in nature and biodegradable (Raman et al 1996 and Ramaswamy 1999). Industrial effluent has been widely used for agricultural purposes nowadays. Researchers have tried industrial effluent from paper mill (Rajanan and Oblisami 1979; Pushpavalli 1990 and Srinivasachari 2000), sugar mil (Zalawadia et al 1997 and Kathiresan et al 1998) and distillery unit (Mohan Rao 1998 and Nagappan et al 1998) and favoured effluent irrigation from the respective industries for different crops. Similarly effluent from industries like textile (Singh et al 2001), tannery (Wilson 1998 and Murthy 1999), rice mill (Pathan and Sahu 1999), sago (Muthuswamy and Jeyabalan 2001), aquaculture (Al-Jaloud et al 1993), treated oil refinery (Aziz et al 1994), steel plant (Sharma and Naik 1999), soap and detergent (Somasekhar and Seetharamaiah 1993) and olive mill (Cox et al 1997) were assessed. These research papers in general advocates the cautious use of effluent (i.e) dilution and irrigation for non-consumable crops. Based on the type of the industry, the effluents may be beneficial or harmful to crop plants (Somasekhar et al 1984). It is estimated that as much as 40 - 50% of water can be reused out of water or effluent discharge of paper mill, iron and steel and thermal power plant (Ramana 1991). Sarikaya and Eroglu (1993) grouped the industries depending on the possibility of irrigation reuse of their wastewater as shown in Table 1.

Industry Group	Possibility of irrigational reuse	Some selected examples
I	Irrigational reuse of effluent is permitted	Beer, Vine, Yeast, Starch, Micro food canning
II	May be permitted on certain conditions	Sugar, Slaughter house, Meat processing, Tanning, Pulp and paper, Textile
III	Not suitable as irrigation water	Paint, Polish, Soap, Pharmaceutical, Metal sulphide, cellulose.

Table 1. Industrial Effluent for Irrigation

Whenever an industry discharges its effluents in a hydrological basin, recycling of wastewater for irrigation has to be encouraged. The decision to use effluents for irrigation is dictated by factors like location of the industry, seasonal increase in irrigational practice, where low flow of surface water coincides with the increase in irrigational water, advanced agricultural practices necessitating supplementary irrigations at various stages of plant growth to derive best advantage, drought condition and rising cost of wastewater treatment. In a tropical and developing country like India, where the water for irrigation is in short supply, above considerations lay emphasis on the priority for an industrial wastewater to be disposed of as irrigation water.

Industries today face the question on how to dispose the effluents. The problem is still more acute for industries, which consume large quantity of water, and in turn lets out huge volume of effluent. Paper industry comes under this category. Paper pervades all walks of human activity and thus pulp and paper industry has been responsible for important technical, social and economic impacts in a country. Basic process of papermaking has gone through a few modifications when compared to other industries, yet water could not be replaced though its use has been reduced to a great extent. The industry continues to utilize large quantities of water right from the stage of washing fibrous raw materials to the drying of paper. In the Indian context, around 200-240 m^3 of water is consumed per tonne of paper and to that extent around 180 - 220 m^3 of effluent is generated. Pulp and paper industries in general are among those highly polluting industries in India and Central Pollution Control Board has identified it as one among the 17 polluting industries for monitoring and regulating the pollution from them.   Considering the industrial growth vis-à-vis the pollution of water resources, even the advanced countries have never tried to curb the industrial growth in spite of insistence of pollution control measures.   In developing countries, particularly those in arid parts of the world, there is a need to develop low cost and low technology methods of acquiring new water supplies for their exploding population while protecting the existing source from pollution. The utilization of industrial effluents for irrigation is an appropriate solution in this context as it involves two main principles – use of soil as a treatment system preventing pollution of the surface water and use of wastewater as continuous or supplementary source of irrigation. In certain cases the dearth of nutrients can also be possibly compensated to a limited extent. Thanks to the effective legislation and implementation of pollution control measures by the State Pollution Control Board. Effluent irrigation programme has been adopted with great vigour by most industries. The present paper throws light on the characteristics of the paper mill effluent water and its suitability for irrigation.

2. Materials and methods

The paper mill located at Pallipalayam in Nammakal district, Tamilnadu was selected for the present study. The effluent treatment plant in paper mill includes primary, secondary and tertiary treatment units. The primary treatment plant includes primary clarifier, vacuum filter and the secondary treatment includes the aerobic lagoon and anaerobic lagoon. The tertiary treatment system consists of masonry baffles, cascades and a bed of blue metal chips. The wastewater collected at various sources is let into the degreasing tank where oil and grease are removed by an oil skimmer or manually. Lime is added in the inlet channel itself to attain a pH of 8.0. The water from the degreasing tank is collected in a collection tank and pumped to first aerator tank where a floating type aerator is arranged. In the collection tank, the pH and temperature of water are stabilized. There are two fixed aerators in second aerator tank. Thus water is exposed to atmospheric air continuously and the BOD of water is much reduced. Nutrients such as urea (30 kg/day) and di ammonium phosphate (20 kg/day) are added in solution continuously at the aerator tank. The over flow water from second aerator tank is let into the clarifier. The sludge formed at the bottom of the clarifier is recirculated till it reaches 3000 rpm. If it exceeds this value, it is sent to the sludge drying beds. The clear overflow from the clarifier (i.e. treated effluent) is pumped to the irrigation fields. The treated effluent water from the pulp mill was pumped into the anaerobic lagoon located 3 km away from the paper unit. After reduction of BOD level, the effluent water is supplied to the farmers through high-density polythene pipes. There were four wastewater streams coming from the industry and let in for irrigation after the treatment and hence four sampling stations (E1, E2, E3 & E4) were identified for periodic sampling. (Fig 1). Quarterly sampling was done for three years. New 1-liter polyethylene bottles were used for sample collection and preservations. The characteristics of the effluent was assessed by the chemical analysis of effluent waters as per the standard procedures (APHA,1995) and the suitability is evaluated for salinity hazard, sodicity hazard, alkalinity and toxicity using parameters such as SAR, Kellys ratio, USSL classification etc and the formulae is given in Table 2.

Parameter	Author	Formula
ElectricalConductivity (dS/m)	USSL (1954)	EC value
Sodium Percent	Eaton (1950)	$\dfrac{Na \quad *100}{(Ca+Mg+Na+K)}$
SodiumAdsorption Ratio (SAR)	USSL (1954)	$\dfrac{Na}{\sqrt{(Ca+Mg)/2}}$
Kelley's Ratio	Kelley (1940)	$\dfrac{Na}{(Ca+Mg)}$
Residual Sodium Carbonate (RSC)	Eaton (1950)	$(CO_3 + HCO_3) - (Ca + Mg)$
Magnesium hazard	Paliwal (1972)	$Mg / (Ca + Mg)$
Chloride concentration	Ayers and Branson (1975)	Chloride concentration in meq/l

Table 2. Parameters for the suitability of effluents for irrigation

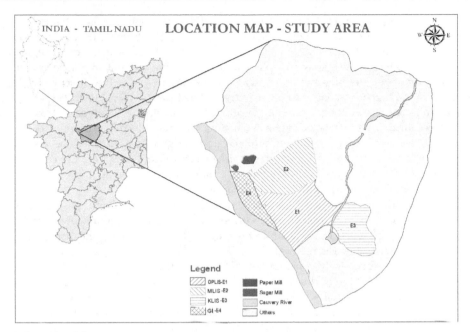

Fig. 1. Location Map

3. Results and discussion

3.1 Effluent characteristics

The average values of the effluent quality collected at different locations are given in the Table 3.

Parameters	E1	E2	E3	E4	Irrigation standards (BIS 1988)
pH	7.99	7.85	7.70	7.91	5.5 - 9.0
EC(dS/m)	1.65	1.51	1.66	1.68	
BOD(mg/l)	48.00	12.00	9.60	36.00	100
DO(mg/l)	0.80	2.90	2.70	0.90	
COD(mg/l)	318	73	81	281	250
Ca(mg/l)	115.11	120.75	107.78	109.56	
Mg(mg/l)	51.33	42.38	49.11	57.22	
Na(mg/l)	147.00	134.75	177.89	144.22	
K(mg/l)	76.11	22.25	20.44	21.78	
HCO$_3$(mg/l)	295.00	281.50	307.11	277.89	
SO$_4$(mg/l)	69.44	152.88	115.33	87.89	1000
Cl(mg/l)	306.49	283.63	355.11	353.22	600
Total N(mg/l)	1.90	1.70	2.01	1.80	
Total P(mg/l)	0.09	0.03	0.09	0.01	

Table 3. Characteristics of effluents

The colour of the effluent let out for irrigation for different schemes was of light brown in colour and the effluent had phenolic smell. Srinivachari (2000) reported that the colouring body present in the wastewater was organic in nature and contain wood extractives lignin and degradation products. The unpleasant phenolic odour of the effluent might be due to the presence of hydrogen sulphide and other organic sulphides present in the effluent, which was evident from the higher content of sulphate in the effluent sample (69.44 to 152.88 mg/l). The effluents collected from different sampling locations are alkaline (pH ranged of from 7.70 to 7.99, which confirms that they are alkaline in reaction. This might be due to addition of sodium compounds viz., caustic soda, sodium sulphate and sodium carbonate for the cooking of chopped raw material in the Kraft pulp process. The electrical conductivity (EC) of the effluent ranged from 1.51 to 1.68 dS/m. This high EC could be attributed to the use of various inorganic chemicals at various stages of paper manufacturing.

The BOD values ranged from 9.60 to 48 mg/l which is well within the permissible limit of 100 mg/l (BIS, 1984). Since the organic wastes in spent Kraft pulping liquor are burned and recovered for energy, they do not contribute to the mill effluent BOD and hence the BOD loading in kraft mill effluent is relatively low. The BOD of the effluent at E1 and E4 were 48 mg/l and 36 mg/l respectively, whereas the BOD of the effluent samples at E2 and E3 were well within 30 mg/l and can be drained into the river (BIS, 1984). The effluents from E2 and E3 are allowed to go to anaerobic lagoon and the reduction in BOD could be achieved by the prolonged anaerobic degradation of dissolved organic matter by biological oxidation. The amount of dissolved oxygen was less in E1 (0.8 mg/l) and E4 (0.9 mg/l). The DO in other samples varied between 1.3-2.9 mg/l. The COD was above 250 mg/l in E1 (381 mg/l) and E4 (281 mg/l) whereas the effluent samples from E2 and E3 are well within the permissible limit.

The average values of cations calcium, magnesium, sodium and potassium content in the treated mill effluent collected from different locations ranged from 107.78 to 120.75 mg/l , 42.38 to 57.22 mg/l. 134.75 to 177.89 mg/l and 20.44 – 76.11 mg/l respectively. The average values of anions viz, bicarbonate, sulphate and chloride content of the effluents in the study area ranged from 277.50 to 334.00 mg/l, 69.44 mg/l to 152.88 mg/l from 283.63 mg/l to 355.11 mg/l respectively. The high sulphate content in the paper mill effluent has been reported by Sreenivasalu (1999). This high chloride content may be due to the chemicals used in bleaching. The present study showed that the effluents contain relatively higher concentrations of cations and anions. The perusal of effluent characteristics given in Table 3 shows that the effluents are well within the permissible limits according to Indian standards IS 2490 part – I 1981 (BIS -1988).

It could also be observed from the Table 3 that the concentration of total nitrogen ranges from 1.3 to 2.01 mg/l and total phosphorous ranges from 0.01 to 0.09 mg/l. This indicates that a major inorganic nutrient in the effluent collected from different sampling sites was of low concentration. This might be due to the little usage of nitrogen and phosphorous containing chemicals in papermaking. Javireen (1991) also stated that the concentration of inorganic nitrogen is low and hence biological methods for the nitrogen removal are not required.

3.2 Suitability of effluents for irrigation

The characteristics of the effluent was compared with the tolerance limit for the industrial effluents for irrigation as given by BIS (1988) (Table 4). Various specifications have been proposed from time to time by different workers to assess the suitability of irrigation water. The guidelines for acceptable salinity and minor element levels in effluent water follow those of normal irrigation water (Ayers & Westcot, 1985). For the present study the characteristics of irrigation water that has been recognized to be the most important were salinity, sodicity, alkalinity and toxicity. Thus the suitability of effluent based on various quality criteria is given in Table 4.

Parameters	E1	E2	E3	E4	Permissible limits	
Electrical Conductivity (dS/m)	1.51	1.65	1.66	1.68	< 0.25	Low
					0.25 – 7.5	Medium
					7.5 – 2.25	High
					> 2.25	Very High
Sodium Adsorption Ratio	2.69	2.86	3.56	2.78	< 3	Low
					3 – 9	Medium
					9 – 26	High
					> 26	Very High
Kelley's Ratio	0.62	0.64	0.82	0.62	< 1	Good
					1 – 2	Marginal
					> 2	Unsuitable
Sodium`Percentage	36.76	34.92	43.76	36.89	< 60 %	Good
					60 – 75%	Permissible
					> 75%	Unsuitable
Residual Sodium Carbonate (meq/l)	-4.80	-4.89	-4.38	5.28	< 1.25	Good
					1.25 – 2.5	Suitable
					> 2.5	Unsuitable
Mg Hazard (%)	36.64	42.36	42.89	46.26	< 50	Good
					> 50	Unsuitable
Chloride concentration (meq/l)	7.91	8.55	9.98	9.85	< 4	Excellent
					4 – 7	Good
					7 – 12	Permissible
					12 – 20	Doubtful
					> 20	Unsuitable

Table 4. Suitability of effluent based on quality parameters

3.2.1 Salinity hazard

The salinity of the irrigation water was evaluated based on electrical conductivity (EC) measurement, The EC of the effluents collected from the different schemes ranged from 1.51 to 1.68 dS/m (Table 1), Saline water can be used for irrigation with suitable amendments and better management practices like leaching with rainfall and low salinity pre plant irrigation of 150 mm or more (Ayars et.al 1993). Gupta (1990) has shown that EC upto 10 dS/m could be utilized for growing tolerant crops on well drained soils were annual rainfall is more than 400 mm. This is evidence of the fact that the actual suitability of given water for irrigation greatly depends on the relative need and economic benefit compared to other alternatives and on the specific condition of use. Based on the USSL classification the effluent fall under 'good to permissible' category with high salinity and low sodium hazard. (Fig 2 & 3). This indicates that the effluents had more salt concentration and their accumulation on the soil affects infiltration and plant growth in the long run.

Best management practices which are site specific needs to be adopted. The farmer of the region also needs to be educated to adopt suitable practices in the available situation. To avoid salt accumulation leaching and drainage has to be done properly. If necessary subsurface drainage could be installed.   Earlier studies have indicated that through drip irrigation the marginal water could be utilized efficiently but care should be taken to prevent the clogging of drips. In case of drip irrigation more area also be irrigated with the effluent water. Selection of crop plays a major role in successful effluent irrigation schemes. Effluent irrigation is encouraged with respect to non consumable crops and landscapping and in some cases with the fodder crops. Another aspect is the crop tolerance and crop rotation. Sugarcane is a tolerant crop and responds well in the effluent irrigation. Reddy et al (1981) has reported sugarcane cultivation with paper mill effluent at Ralgada region of Orissa, India. Sugarcane cultivation using tannery effluent was reported by Kathiresan et al (1998).The Sugarcane mill effluent and distillery effluent for sugarcane cultivation was earlier reported by Nemade (2002).

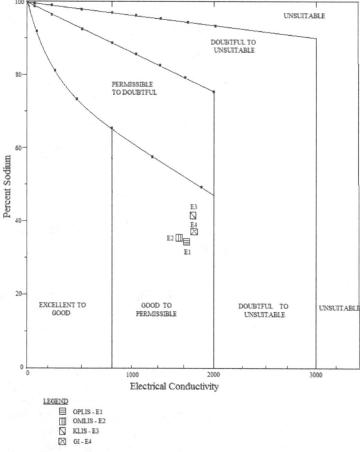

Fig. 2. USSL Plot (Wilcox 1984)

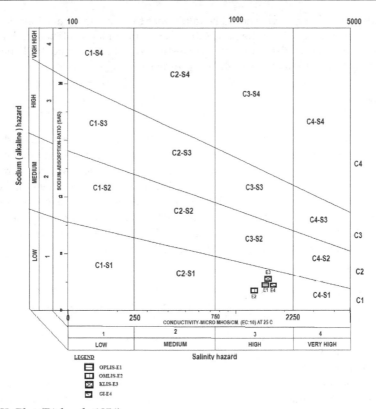

Fig. 3. USSL Plot (Richards 1954)

3.2.2 Sodicity hazard

The sodicity of the irrigation water was determined by sodium percent and sodium adsorption ratio (SAR). The sodium concentration is important in classifying irrigation waters because sodium by the process of base exchange may replace calcium in the soil and thereby reduce the permeability of the soil water. If this process continues it affects the plant growth. From the Table 3, it could be seen that the percent sodium was less than 60, which implies that the paper effluent comes under the category 'good'. Based on SAR, it comes under the low category, as the effluent water collected from different sampling locations registered SAR values less than the 10 and ranged from 2.69 to 3.56.The Kelly's ratio of the effluents collected at all the sample points ranged from 0.62 to 0.82, which is well below unity indicating that the water is free from sodicity hazard.

3.2.3 Alkalinity

The RSC values were negative (Table 3) which means it comes under the category 'Good'. Although CO_3 is much more toxic than bicarbonate and nature of magnesium ions is different than calcium ions, the two anions and cations have been paired together assuming similar effect. In absence of carbonate ions, the bicarbonate does not precipitate Mg. In such cases, Gupta (1984) suggested that alkalinity should be measured using $HCO_3 - Ca^{2+}$ and called as residual carbonate. Care should be taken that these components does not build up in groundwater reserve. Water conservation techniques like artificial recharging the groundwater reserve using the rainwater harvesting structures would be beneficial and helps for the long term sustainability of the programme.

3.2.4 Toxicity

3.2.4.1 Magnesium hazard

The term Magnesium hazard has been used by Paliwal (1972) to evaluate the hazardous effects of Mg to irrigation water and stated that Mg hazard is likely to be developed in the soil when the Mg ratio exceed 0.50. The degree of hazardous effects would increase with increase of Mg / Ca ratio. However, the harmful effect of Mg of irrigation water on the soil is likely to be reduced by the release of Ca on dissolution of $CaCO_3$ if present in the soil (Gupta, 1994). From the Table it could be observed that Mg/Ca ratio in all the effluent samples was less than 0.50 indicating that there was no magnesium hazard in utilizing the mixed effluents for irrigation. Irrigation waters with 0.50 Mg ratio can be safely used for irrigation.

3.2.4.2 Chloride toxicity

Ayers and Brason (1975) proposed the chloride tolerance in waters to be used as an indicator for irrigation purpose. The chloride concentration in the effluents let out for irrigation ranged from 7.91 to 9.90 meq/l which implies that the effluents are permissible for irrigation (Table 3). Gupta (1989) resorted that if the chloride concentration ranges between 4-10 then suitable amendment has to be applied. This again suggests the required amendment and management practice for the long term usage for irrigation. Application of amendments like gypsum and press mud will help in a major way in maintaining the soil health. Alfred et al 1998 identified the press mud and effluent treatment plant sludge as effective ameliorants to overcome salinity due to paper mill effluents. this is because they act as a chealating agent in binding the divalent cation present in paper and pulp effluent.

4. Conclusion

From the aforesaid discussion and interpretation of the chemical quality of the effluent let out for irrigation, it can be concluded that in general the treated paper mill effluents are found to be well within the permissible limits for irrigation. The study further reveals that the quality of the treated effluent falls under the category of 'permissible to suitable' emphasizing the need for better management practices for the long term sustainability of the effluent irrigation programme. The effluents from agro based industries, which use agriculture products as the raw material, and industries, which involved in processing of agricultural products are not detrimental to the environment since the wastes are organic in nature and biodegradable. It is a fact that the suitability of the effluent water for irrigation and the implementation by the farmers is dependant upon the condition of use including crop, climate, soil, irrigation methods and management practices, rather than the water quality classification. Hence better irrigation methods and proper monitoring system will help to minimize the risks and maximize the benefits from this resource. Watershed based impact assessment and conservation strategies needs to be developed, and adopted with the stakeholders participation.

5. References

Abul Basheer M Shahalam (1989), 'Wastewater effluent Vs. Safety in its reuse : State – of – art', J. of Environ. studies, pp. 35 – 42.

Alazaba A.A. (1998), 'Necessity for modification of management parameters when using low quality water', Agricultural Water management, Vol. 36, No.3, pp. 201 – 211

Alfred Sahaya, Udayasooriyan R.C. and Ramaswami P.P. (1998), 'Impact of treated paper mill effluent irrigation on rice soil eco system', Water World'98, Proc. of National Seminar on wastewater for irrigation, pp. 50 – 59.

APHA (1995), 'Standard methods for the examination of water and Wastewater', 19th edition, APHA, AWWA, WPCF, New York, p. 1193

Asano Takashi (2002), 'Water from Wastewater the dependable Water Resource', The 2001 Stockholm Water Prize Laureate Lecture, Wat. Sci.Tech., Vol. 45, No. 8, pp. 23 – 33.

Ayars J.E., Hutmacher R.B., Schoneman R.A., Vail S.S. and Pflaun T. (1993), 'Long term use of saline water for irrigation', Irrigation Sci., Vol. 14, pp. 27 – 34.

Ayers R.S. and Branson R.L. 1979. 'Guidelines for interpretation of water quality for agriculture' Univ. of California, Extention Mimeographed, p. 13.

Ayers R.S. and Westcot D.W. (1985), 'Water quality for agriculture', Irrigation and Drainage Paper No. 29, Revision I, F.A.O., Rome.

Bajwa M.S. (1997), 'Water quality', In : National Water Policy, Agricultural Scientist Perceptions, Eds. Randhawa and Sarma P.B.S., Pub. No. 7, NAAS, NewDelhi, India, pp. 260 – 268.

Barbagallo S., Cirelli G.L. and Indelicato S. (2001), 'Wastewater reuse in Italy', Wat. Sci. Tech., Vol. 43, No. 10, pp. 43 – 50.

Bureau of Indian standards, 1984. Tolerance limits for industrial effluent discharged on land for irrigation purposes Is: 2490, 1984.

Cox L., Celis R., Hermosin M.C., Berker A. and Cornejo J. (1997), 'Porocity and herbicide leaching in soils amended with Olive mill wastewater,' Agri. Eco. Syst. Environ., Vol. 65, No. 2, pp. 151 – 162.

Eaton E.M. 1950. Significance of carbonation in irrigation waters, Soil Sci., Vol. 69, pp. 123 – 133.

Franco J.A., Abrosqueta J.M., Hernansoez A. and Moreno F. (2000), 'Water balance in a young almond orchard under drip irrigation with water of low quality', Agri. Wat. Mgmt., Vol. 43, No. 1, pp.75 – 95

Gupta D.C. (1989), 'Irrigational suitability of surface waters for agricultural development of the area around Mandu, District, Dhar, Madhya Pradesh, India, J. of Applied hydrology, Vol. II, No.2, pp. 63 – 71.

Gupta I.C. (1990), 'Use of saline water in Agriculture. A study of arid and semi arid zones of India', Revised ed., Oxford & IBH publishing Co., New Delhi, p. 308.

Gupta I.C. (1994), 'Reassessment of standards for industrial effluents discharge on land for irrigation', Current agriculture, Vol. 18, No.1 – 2, pp. 65 – 70.

Hussain I., Rasheed L., Hanjra M.A., Marikar F. and Vanderhoek W. (2001), 'A frame work for analysing socio–economic health and environmental impacts of wastewater use in agriculture in developing countries', Working paper 26. Colombo, Srilanka, IWMI.

Jarvinen, R. 1997 , ' Nitrogen in the effluent of the pulp and paper industry . Water Science and Technology Vol 35 : 139-145.

Kelley W.P., Brown S.M. and Liebig G.F.Jr. 1940. 'Chemical effects of saline irrigation waters on soils', Soil Sci., Vol. 49, pp. 95 – 107.

Larson T.E. and Boswell A.R. (1942), 'Calcium carbonate saturation indices and alkalinity interpretation', Am. Wat. Works Assn. J., Vol. 34, pp. 1667 – 1687.

Paliwal K.V. 1972, 'Irrigation with Saline water', I.A.R.I., Monograph No. 2 (New Series), New Delhi, pp. 198.

Richards L.A. 1954. 'Diagnosis and improvement of saline and alkaline soils', Agri. Hand book. No. 60, U.S. Dept. Agri., Washington D.C., pp. 160.

Shelef G. and Azov Y. (1996), 'The coming era of intensive wastewater reuse in the Mediterranean region', Wat. Sci. Tech., Vol. 33, No.10 – 11, pp. 115 – 125.

Sreenivasalu A., Sundaram E.V. and Reddy M.K. 1999. 'Correlation of physical – chemical characteristics of ITC paper mill', Indian J. Environ. Prot., Vol. 19, No. 100, pp. 767 – 770

Srinivasachari M., Dakshinamoorthy M. and Arunachalam G. (2000), 'Accumulation and availability of Zn, Cu, Mn and Fe in soils polluted with paper mill wastewater', Madras Agric. J., Vol. 87, No. 4 – 7, pp. 237 – 240.

Srinivasachari Matli, Dakshinamoorthy M. and Arunachalam G. (1999), 'Studies on the influence of paper mill effluents on the yield, availability and uptake of nutrients in rice', J. Indian Soc. Soil. Sci., Vol. 47, No. 2, pp. 276 – 280.

Tanji K.K. (1997), 'Irrigation with marginal quality water: Issues', J. of Irrigation and Drainage Engg., Vol. 123, No. 3, pp. 165 – 169.

Tsagarakis K.P., Mara D.D. and Angelakis A.N. (2001), 'Wastewater management in Greece: experience and lessons for developing countries', Wat. Sci. Tech., Vol. 44, No. 6, pp. 163 – 172.

Videla Sand and Diez C. (1997), 'Experiences of wastewater treatment in Chilean forest industry', Wat. Sci. Tech., Vol. 35, No. 2 – 3, pp. 221–226.

Wilcox L.V. (1948), 'The quality of water for irrigation use', U.S.D.A. Tech. Bull. 962, Washington D.C., p. 40.

Zalawadia N.M., Raman S. and Patil R.G. (1997), 'Influence of diluted spent wash of sugar industries application on yield and nutrient uptake of sugarcane and changes in soil properties', J. Indian Soc. Soil Sci., Vol. 45, No. 4, pp. 767 – 769.

11

Risks for Human Health of Using Wastewater for Turf Grass Irrigation

Pilar Mañas, Elena Castro and Jorge de las Heras
Centro Regional de Estudios del Agua (CREA), Universidad de Castilla-La Mancha
Spain

1. Introduction

In recent years, continuous population growth in most Mediterranean countries such as Spain has caused an increase in consumption of existing water resources. This population increase has not only increased freshwater demand but has also increased the volume of wastewater generated (Quian and Mecham, 2005). Thus, there is an urgent need to conserve and protect freshwater and to utilize the wastewater generated (Gan et al., 2006). Using treated wastewater for agricultural irrigation helps to alleviate demand of scarce potable water and groundwater resources (Angin et al., 2005). Water scarcity and water pollution pose a critical challenge in many developing countries and it is difficult for authorities to manage water supplies and wastewater (Chizuru et al, *2005)*. There is a significant absence of legislation in the EU in controlling wastewater reuse in agriculture. Currently there are no international standards except for the Worldwide Health Organization Guidelines, which are starting points for setting water quality standards, including microbiological standards (Campos, 2008). The World Health Organization (WHO), the US Environmental Protection Agency (USEPA) and the World Bank have reviewed the public health aspects of crop irrigation with domestic wastewater and have made recommendations for the microbiological quality of treated wastewaters used for this purpose (Shuval et al. 1989; WHO 1989; USEPA 1992). A Spanish law (R.D 1620/2007) establishes limits on the use of wastewater depending on the type of application in which the irrigation of green spaces (sports fields, parks) and gardens (private and public) have been considered. Wastewater can be a resource but may present a hazard at the same time (WHO, 2006). Proper wastewater reuse may offer solutions for meeting water resource needs. The fundamental precondition for water reuse is that applications will not cause unacceptable public risks (Chizuru et al, *2005)*. According to Dr. John Sheaffer, the president of Sheaffer International, Ltd., (McKenzie, 2005) "Wastewater can be viewed as a resource, fresh water containing plant nutrients (nitrogen, phosphorus, and potassium). In the groundwater, these nutrients are a pollutant, but on a growing crop or turf, they are a resource. When wastewater is reused, it is not available to pollute the groundwater supply."

A common type of recycled water is water that has been reclaimed from municipal wastewater (sewage). Different, specific parameters must be analyzed depending on the origin of the wastewater and the intended use. Simple parameters such as salinity, *E. coli*, turbidity, TSS, organic matter, DOC and other N- and P-related variables offer useful information depending on the final use of the reclaimed water (Salgot et al, 2006).

The quality of irrigation water is of particular importance in arid zones where extremes in temperature and low relative humidity result in high evaporation rates, with consequent deposition of salt that tends to accumulate in the soil profile (Pescod, 1992).

The use of wastewater irrigation in turf grass could be an alternative to drinking water irrigation since several studies confirm that treated wastewater can be used for turf grass irrigation with a minimal environmental impact (Wu et al., 1996; Barton et al., 2005; Menzel and Broomhall, 2005; Lockett, 2008; Castro et al, 2011).

Otherwise, wastewater treatment plants (WWTP) represent a common source of odor emissions. Odor generated by wastewater using a sprinkler system to irrigate a park or public garden may condition use because of the nuisance it causes to the population. There are many studies on this topic (Lawrence & Tan, 1990; Capelli et al, 2009; Cheng et al, 2009; Bo et al, 2011), but it is difficult to avoid the problem unless effective disinfection treatment is used. The selection of which method works best depends on the concentration of the odor causing compounds, the air flow rate, available land area for the system, capital budget and discharge limitations for wastewater from the system.

Coliform bacteria are organisms present in the environment and in the feces of all warm-blooded animals and humans. They will not likely cause illness, but their presence in drinking water indicates that disease-causing organisms (pathogens) could be in the water system. Most pathogens that can contaminate water supplies come from human or animal feces. There are three different groups of coliform bacteria, each with a different level of risk: total coliform, faecal coliform, and E. coli. The total coliform group is a large group of different kinds of bacteria. Faecal coliforms are types of coliforms that mostly exist in feces and E. coli is a sub-group of faecal coliforms. Most E. coli bacteria are harmless and are found in great quantities in the intestines of people and warm-blooded animals. Some strains, however, can cause illness. This is the case of Enterohaemorrhagic E. coli (EHEC), recently occurred in Germany. It can cause diarrhea, severe stomach cramps and fever (Davis, 2011). The presence of E. coli in a drinking water sample almost always indicates recent fecal contamination, meaning there is a greater risk that pathogens are present (DOH, 2007). Total coliform bacteria are commonly found in the environment (e.g., soil or vegetation) and are generally harmless. If only total coliform bacteria are detected in drinking water, the source is probably environmental and faecal contamination is not likely. However, if environmental contamination can enter the system, there may also be a way for pathogens to enter the system.

Another test parameter is helminth populations, which are multicellular organisms. The free-living larvae are not usually pathogenic and they have high resistance to adverse environmental conditions and disinfectants. They are well adapted to survive in water systems and in some cases they emerge alive from domestic taps (Campos, 2008). Ascaris (a nematode) is the most common helminth egg in wastewater and sludge. Eggs contained in wastewater are not always infective. To be infective they need to develop larva, for which a certain temperature and moisture are required (26° C and 1 month in laboratory conditions). These conditions are usually found in soil or crops where eggs, deposited through irrigation with wastewater or sludge, can develop larva in 10 days. Helminth ova (or Helminth eggs) can live in water, soil, and crops for several months/years (Feachem et al., 1983 cited by Jimenez, 2007). In previous studies, sewage sludge and wastewaster from the Wastewater Treatment Plant (WWTP) of the city of Albacete was determined to be safe and adequate for agricultural uses (de las Heras et al, 2005; Mañas, 2006; Mañas et al, 2010). However, risks to human health derived from microbial content in this type of water resource have not been well evaluated. Therefore, the main goals of the present study were: to make a replicate of a garden or a public park irrigated with treated wastewater to evaluate the applicability of treated wastewater for turf grass by assessing the physical and chemical effects of continued usage of treated water on the soil and plant over a two-year period and to assess the human health risk and, if possible, to define the availability of wastewater for this use.

2. Materials and methods

2.1 Design of experiment and localization

The field trial was carried out in two square study plots, of 225 m² each, on farmland near the WWTP (Figure 1) of Albacete (165,000 inhabitants) in SE Spain (39° 00′ 57.82″ N; 1° 50′ 50.45″ W). The source of city drinking water was surface water at the time of the study. Two types of water treatments were considered: drinking water (D), considered the control, and treated wastewater (W). The origin of wastewater was mixed: approximately 70% of domestic origin and 30% industrial from two industrial areas. Most of these industries are related to knife manufacturing, automobile replacement parts and a few food producers.

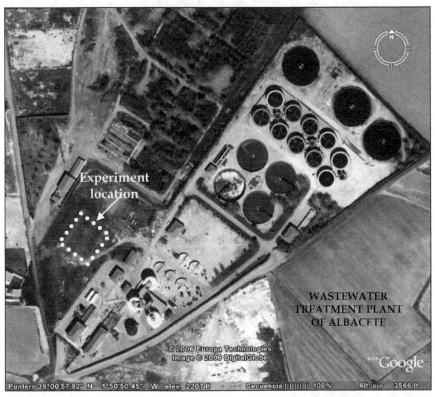

Fig. 1. Location of the experiment in the wastewater treatment plant of Albacete

The city wastewater treatment plant uses trickling filters with an open-air system. The trickling filters consist of a plastic medium over which wastewater flows downward and causes a film of microbial slime to cover the medium bed. Filtered wastewater is poured into a channel that, for years, local farmers have mainly used to irrigate their corn and winter cereal cultivations using a flooding irrigation system.

The study was conducted May 2007-August 2008 (during 466 days after planting). Rainfall (mm), monthly ET_0 (mm), mean average temperature (°C) and sunlight (MJ m^{-2}) during the study period are shown in Figures 2 and 3.

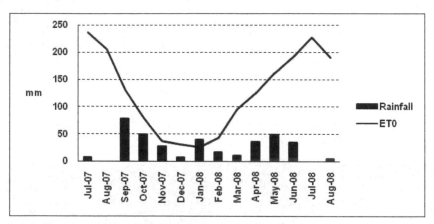

Fig. 2. Rainfall(mm) and ET_0 (mm) during the study period.

Throughout this period the mean minimum, mean maximum and mean average temperatures were 8.1°C, 23.0°C and 15.5°C, respectively, total precipitation was 373.8 mm, average solar radiation was 20.3 MJ m^{-2} and mean daily sunshine was 10.9 hours.

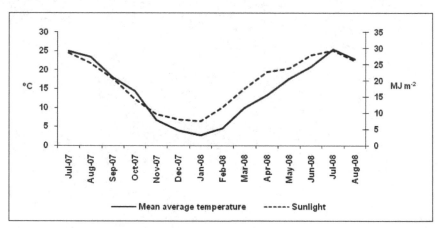

Fig. 3. Mean average temperature (°C) and sunlight (MJ m^{-2}) during the study period.

At the beginning of the first year, a sprinkler irrigation system was installed in each plot corner (15 m x 15 m) and the land was prepared with suitable farm machinery in order to apply 12 mm of mulch. One of the plots (control) was irrigated with drinking water and the other received treated wastewater (Figure 4). Each plot was divided into four sites (replicates).

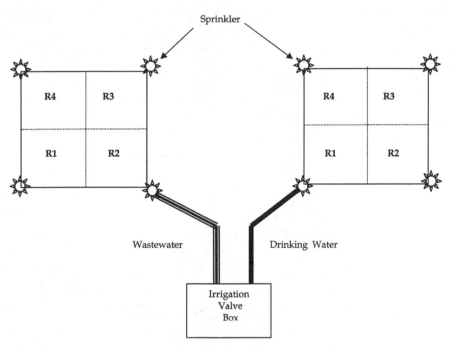

Fig. 4. Sketch of the experimental design. (R1, R2, R3 and R4: Replicates 1, 2, 3 and 4).

2.2 Sowing and Irrigation

In May, 2007, 35 g m^{-2} of grass seed mixture was planted. The composition of the mixture was 65% of *Festuca Arundinacea* Kilimanjaro, 20% *Festuca Arundinacea* Arid 3,5% *Poa Pratense* Conni and 10% *Lolium perenne* Delaware. Weed control was performed using 2-methyl-4-chlorophenoxyacetic acid (MCPA). The plots were irrigated regularly to avoid drought stress. To calculate the water needs for this crop, we followed the methodology proposed by the FAO (Doorenbos & Pruitt, 1984), which calculates the Reference Evapotranspiration (ETo) using the Penman-Monteith method and applies the crop coefficient (Kc). Since the ETo and Kc values are known, Evapotranspiration and therefore water needs could be calculated. A total of 1688 mm of both types of water were distributed in irrigation. During the winter season (October 2007-May 2008), the crop was not irrigated because in this period the crop ceased growth and the risk of frost could damage both the crop and the irrigation system.

2.3 Sampling
2.3.1 Water

Treated wastewater and drinking water were tested ten times while the crop was in the field. The chemical composition of irrigation water (D and W) and some microbiological parameters (total coliforms, faecal coliforms, *Salmonella s.p* faecal streptococci and sulphite-reducing clostridia) were determined and recorded (Table 1). In drinking water, BOD$_5$, TSS and microbiological parameters were not tested. The composition of the two types of irrigation water does not vary significantly over the study period, so the mean values ± standard deviation for the chemical properties have been presented in both cases except for helminth eggs.

PARAMETER	UNITS	DRINKING WATER	TREATED WASTEWATER
		Average±sd	Average±sd
COD	mg l^{-1} O_2	<10	125±86
BOD$_5$	mg l^{-1} O_2	Not detected	31.7±21.3
TSS	mg l^{-1}	Not detected	23±18
pH		8.1±0.2 a	7.9±0.3 a
E.C.	μS cm^{-1}	815±80 a	1759±237 b
Total water hardness	mg l^{-1}	488±70	NT
Dissolved oxygen	mg $l^{-1}O_2$	NT	2.5±1.7
Phosphorus	mg l^{-1}	NT	4.2±1.4
N-Kjeldahl	mg l^{-1}	NT	58.3±31.7
N-Ammoniacal	mg l^{-1}	NT	31.6±10.7
Nitrite	mg l^{-1}	<0.01	4.4±1.4
Nitrate	mg l^{-1}	1.7±0.8 a	4.8±0.9 b
Sulphate	mg l^{-1}	336±66 a	344±72 a
Carbonate	mg l^{-1}	NT	46.3±15.5
Bicarbonate	mg l^{-1}	NT	332±43
Chloride	mg l^{-1}	28.6±4.5	242±7
IC	mg l^{-1}	26.9±0.4 a	83.6±42.5 b
TC	mg l^{-1}	29.5±0.9 a	104±50 b
TOC	mg l^{-1}	2.5±0.5 a	20.2±7.5 b
SAR		0.2±0.1 a	2.5± 1.1b
Na	mg l^{-1}	13.6±3.0 a	170±98 b
K	mg l^{-1}	2.1±0.08 a	43.6±54.3 b
Ca	mg l^{-1}	112±22 a	138±50 a
Mg	mg l^{-1}	50.2±8.6 a	56.1±24.4 a
Zn	mg l^{-1}	<0.24	0.5±0.4
Al	μg l^{-1}	16.2±4.5 a	2786±1582 b
Cu	μg l^{-1}	3.9±3.2 a	200±163 b
Fe	μg l^{-1}	23.6±14.4 a	2120±1249 b
Pb	μg l^{-1}	<7.5	31.1±22.4
Cd	μg l^{-1}	<1	2.1±2.8
Cr	μg l^{-1}	<5	30.1±20.4
Mn	μg l^{-1}	8.2±6.8 a	62.3±66.8 b
Ni	μg l^{-1}	<10	54.9±60.2
As	μg l^{-1}	<10	<10
Se	μg l^{-1}	<5	<5
B	μg l^{-1}	0.03±0.02 a	0.3±0.1 b
Hg	μg l^{-1}	<1	1.9±1.4
Total coliforms	cfu 100 ml^{-1}	NT	1.7 104 ± 3.1 104
Faecal coliforms	cfu 100 ml^{-1}	NT	4.1 103 ± 7.7 103
Salmonella sp	cfu 100 ml^{-1}	NT	Not detected
Faecal streptococci	cfu 100 ml^{-1}	NT	3.1 103 ± 3.0 103
Sulphite-reducing clostridia	cfu 100 ml^{-1}	NT	2.4 103 ± 1.5 103
Helminths eggs	Eggs 10 l^{-1}	NT	6

Table 1. Chemical composition of irrigation water. COD: Chemical Oxygen Demand; BOD$_5$: Biological Oxygen Demand, Five-Day; TSS: Total Suspended Solids; EC: Electrical Conductivity; IC: Inorganic Carbon; TC: Total Carbon; TOC: Total Organic Carbon;SAR: Sodium Absorption Ratio. NT: Not Tested. Different letters mean significant differences at $p<0.05$, according to Fisher's LSD test. (n=10).

In addition, in order to describe chemical water data, we constructed Piper diagrams for D and W water (Figure 5). Microorganism presence in drinking water was not analyzed because it was chlorinated. In addition, treated wastewater was tested once in order to analyze for odor (APHA, 1998).

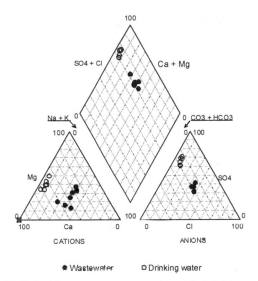

Fig. 5. Piper diagrams for irrigation water: wastewater and drinking water.

2.3.2 Soil

Before sowing a 25 cm deep soil sample from ten random points in the test plots was collected with a hand auger and three replicates were analyzed (n=3). At the end of the study (two months after suspending irrigation) new samples were collected at two different depths: 0-25 cm and 25-40 cm. In this case, soil samples were collected from ten points at random inside each of the four test plots and three repetitions of each were analyzed (S.A.F., 2005).

The original soil (Table 2) was basic, slightly saline (Cros, 1983), with a medium level of chloride, sulphate, organic matter and total nitrogen (Yañez, 1989). According to the C:N ratio, there was high nitrogen liberation (Quemener, 1985; Guigou et al. 1989). The amount of P, K, Ca and Mg was very high (Yañez, 1989). The total carbonate percentage was high but sodium content was low. The K:Mg and Ca:Mg ratios had adequate values (Yañez, 1989).

2.3.3 Grass

Height (cm) and phytomass growth (kg ha^{-1}) in wet weight were recorded six times during the crop season (summer-autumn 2007 and spring-summer 2008) and grass was mowed twice according table 3.

In winter 2007-2008 no plant samples were collected because crop growth was very slow and no differences had been observed. All measurements were collected randomly for one plant in each plot replicate (n=4).

PARAMETERS	UNITS	May 2007 Original soil 0-25 cm
Sand	%	27
Silt	%	29
Clay	%	45
pH		8.12
E.C	mmhos cm^{-1}	0.69
Chloride	(mg gypsum) (100 g soil)$^{-1}$	27
Sulphate	(mg gypsum) (100 g soil)$^{-1}$	52
Organic matter	%	2.7
Total N	%	0.18
C:N Ratio		9
Nitric N	mg kg^{-1}	47
Asimilable P	mg kg^{-1}	66
Assimilable K	meq 100g^{-1}	2.08
Assimilable Na	meq 100g^{-1}	0.84
Assimilable Ca	meq 100g^{-1}	24.51
Assimilable Mg	meq 100g^{-1}	5.40
Assimilable B	mg kg^{-1}	2.53
K:Mg Ratio		0.39
Ca:Mg Ratio		5

Table 2. Chemical characteristics of the original soil before sowing at 0-25 cm.

Date	Event
18/05/07	Sowing
15/07/07	Sampling
26/09/07	Sampling
24/10/07	Sampling
26/05/08	Mowing
02/06/08	Sampling
22/06/08	Sampling
26/06/08	Mowing
31/07/08	Sampling

Table 3. Dates of mowing and height and phytomass growth sampling events.

To get height and phytomass yields, a grass height meter developed by NMI (Nutrient Managament Institute) and distributed by Eijkelkamp was used to take measurements.

Following the NMI method, thirty measurements from each subplot were recorded to get an average value. Height was derived from the average of four replicate values. The NMI method determines phytomass (kg ha^{-1}) by indirect measurements according to the equation (1):

$$\text{Phytomass (kg ha}^{-1}) = 168.24 \times \text{Height (cm)} + 813.19 \tag{1}$$

Therefore, the height value from each replicate (n=4) was used in this equation in order to determine phytomass. Finally, at the end of the last year foliar tissue samples from each treatment (n=4) were collected to determine chemical parameters such N, P, K, Ca, Mg, Na, Fe, Cu, Mn, Zn and Al.

Microbiological parameters (total coliforms, faecal coliforms, *Salmonella sp.* faecal streptococci and sulphite-reducing Clostridia in foliar tissue were tested eight times from July, 2007 to August, 2008 (Table 4) and a sample of each four-treatment replicate was collected in sterile plastic bags.

Date	Days after planting
18/05/07	0
15/07/07	58
26/09/07	131
24/10/07	159
07/04/08	323
26/05/08	372
22/06/08	399
31/07/08	438
28/08/08	466

Table 4. Dates of microbiological sampling events.

2.4 Analytical methods
2.4.1 Water
Water samples were analyzed with atomic emission (instrument error < 1%) to determine Na and K, atomic absorption spectroscopy (instrumental error <10%) to determine Fe, Al, Cu, Mn, Cd, Cr, Ni, Pb, As and Se and ionic chromatography (instrument error < 10%) to determinate chloride, sulphate and nitrate content.

Following the APHA (1998) method, COD in wastewater was determined by the open reflux method, BOD by 5-Day BOD test, pH by the electrometric method with a previous calibrated CRISON© GLP22 pH meter, TSS by membrane filter technique and EC by conductimetry.

Wastewater odor was tested according to the APHA (1998) method. To ensure precision we used a panel of 10 testers, starting with the most dilute sample first (1:200) to avoid affecting sensitivity with the concentrated sample (1:1). To assess microbiological content in wastewater, the most probable number count methodology was used (APHA, 1998). To determine total coliforms, faecal (F-) coliforms, faecal (F-) streptococci and sulphite-reducing (sr) Clostridia, dissolution from 25 g of fresh sample in tryptone water was prepared. The growth media used were lactose broth for total coliforms and faecal (F-) coliforms, Rothe broth for streptococci and sulphite iron Wilson Blair agar for Clostridia. In every case the incubation period was 24 hours and the temperature was 37°C. To confirm total coliform (gas presence), lactose broth was used, incubating it during 24 hours at 37°C. To confirm faecal (F-) coliforms, E.C. broth was used as a growing medium and the incubation period was 24 hours at 44°C. Finally, to confirm streptococci growth, the medium was Bromocresol purple azide broth with the same incubation period and temperature as for total coliforms.

To determine *Salmonella sp.*, the dissolution was prepared from 25 g of fresh sample in buffered peptone water. The growing medium was Rappaport broth and the incubation period was 24 hours at 42°C. To confirm *Salmonella sp.* presence HE Agar (Hekton Enteric) was used.

Helminth eggs were determining by counting number of eggs per 10 liters of water.

2.4.2 Soil

Soil samples were analyzed with the following techniques: pH with a previously calibrated CRISON© GLP22 pH meter and 1:2 w v-1 suspension of soil in water (Peech, 1965), organic matter using the Walkey Black method (Allison, 1965), electrical conductivity (Bower and Wilconx, 1965), N by the Kjeldahl procedure (Bremmer, 1965) and extractable P (Olsen, 1965). Besides, soil samples (0.5 g dry weight) were prepared for analysis with acid digestion in 4 ml of $HNO3$, 0.25 ml of $H2O2$ and 2 ml of HF and by applying the temperature program according to Milestones'© Cookbook of microwave application notes for MDR technology in order to determine Ca, Mg, and K (atomic emission in Spectr AA 50 Atomic Absorption Spectrometer with SIPS-10, Varian©); Fe, Zn, Al, Cu, Cd, Cr, Ni, Pb, As, and Mn (atomic absorption spectroscopy in Atomic Absorption Spectrometer Spectr. AA 220, Varian©).

2.4.3 Grass

Plant samples (0.5 g dry weight) were prepared for analysis with acid digestion in 6 ml of HNO_3, 1 ml of H_2O_2 and 0.5 ml of HF and we applied the temperature program according to Milestones'© Cookbook of microwave application notes for MDR technology. Next, we used atomic absorption spectroscopy analysis to determine K (d.l.: 0,01 ppm), Zn (d.l.: 0,1 ppm), Mg (d.l.: 0,05 ppm), Ca (d.l.: 0,1 ppm), Al (d.l.: 2 ppm), Cu (d.l.: 0,5 ppm), Fe (d.l.:0,5 ppm) and Mn (d.l.: 0,2 ppm). N in plant samples was analyzed using the Kjeldhal Method and total P was determined by spectrophotometry after acid digestion in HNO_3 and H_2SO_4.

To determine microbiological parameters in foliar tissue, the same technique that in water samples was used.

2.5 Statistical procedures

The experimental design used 2 treatments with 4 replicates of turf grass. Data were subject to Anova treatments and the method used to discriminate among the means was Fisher's least significant difference (LSD) for $p<0.05$. To ensure that data came from a normal distribution, standarized skewness and standarized kurtosis values were checked. Percentage values were transformed by arcsine. The dynamics of microorganisms in turf grass leaves were tested using a simple regression analysis. All statistical calculations were performed with Statgraphics plus 5.1.

3. Results and discussion

3.1 Water

During the study, 1688 mm of both types of water were distributed in irrigation. Important agricultural water quality parameters include a number of specific water properties that are relevant in relation to crop yield and quality, maintenance of soil productivity and environmental protection. These parameters mainly consist of certain physical and chemical characteristics of water (Pescod, 1992). Table 1 shows that electrical conductivity (E.C.), COD (Chemical Oxygen Demand), nitrite, nitrate, bicarbonate, chloride, inorganic carbon (IC), organic total carbon (TOC), total carbon (TC), sodium, potassium, aluminum, copper, iron, lead, cadmium, chrome, manganese, nickel and mercury have significantly higher values in wastewater than in drinking water. In addition, average B content in wastewater is 0.3 mg l⁻¹, which is excellent even for a crop sensitive to boron (Ayers and Westcot, 1987).

The Piper diagram (Figure 5) shows than wastewater is a NaCa-SO$_4$Cl water type and drinking water corresponds to a Ca-SO$_4$ facie. On the other hand, Ayers and Westcott (1987) suggested some guidelines for interpreting water quality for irrigation based on Salinity, SAR, Specific Ion Toxicity (Na, Cl, B and trace elements) and miscellaneous effects, and they defined some degrees of restriction on usage. In Table 5, and according these authors, we can see that drinking water has no degree of restriction on use for sodium, chloride, boron and nitrogen and slight to moderate because of SAR and electrical conductivity. In contrast, treated wastewater has a slight to moderate degree of restriction on use for electrical conductivity, SAR, sodium, and chloride but no degree of restriction on use for boron or nitrogen.

	Drinking Water	Degree of Restriction on Use	Treated Wastewater	Degree of Restriction on Use
Electrical Conductivity (dS cm⁻¹)	0.815	Slight to moderate	1759	Slight to moderate
SAR	0.2	Slight to moderate	2.5	Slight to moderate
Sodium (Na) meq l⁻¹ Sprinkler irrigation	0.59	None	7.39	Slight to moderate
Chloride (Cl) meq l⁻¹ Sprinkler irrigation	0.81	None	6.81	Slight to moderate
Boron (B) mg l⁻¹	0.03	None	0.3	None
Nitrogen (NO$_3$ – N) mg l⁻¹	1.7	None	4.8	None

Table 5. Interpretations of water quality for irrigation according Ayers and Westcot (1987).

The presence of microorganisms in wastewater from highest to lowest was total coliforms (1.7 10⁴ cfu 100 ml⁻¹), faecal coliforms (4.1 10³ cfu 100 ml⁻¹), faecal streptococci (3.1 10³ cfu 100 ml⁻¹) and sulphite-reducing Clostridia (2.4 10³ cfu 100 mL⁻¹). *Salmonella sp.* was not detected in any case.

The pathogens most resistant in the environment are helminth eggs, which in some cases can survive for several years in the soil. Pathogen survival in soil and on different crops can vary. For example, tapeworm eggs can survive in selected environmental media at 20-30 °C for many months in freshwater, sewage and soil and usually less than 30 days in crops, which is the same for *Ascaris* eggs. Nevertheless, this pathogen can survive for years in freshwater, sewage and soil (WHO, 2006). Helminth eggs in wastewater were tested once during the study and 6 eggs 10 l^{-1} were detected. Although the greatest health risks are associated with crops that are eaten raw, this value (6 eggs 10 L^{-1}) exceeds the WHO (1989) recommendation and Spanish legislation (RD 1620/2007) of $\leq$ 1 egg 10 l^{-1}. In this case, this water cannot be used to irrigate grass in a public park.

The threshold odor number (TON) is the greatest dilution of sample with odor-free water yielding a clearly perceptible odor (APHA, 1998). For the odor test, 1 ml of sample (wastewater) was diluted in 200 ml of odor-free water. All ten testers (100%) showed that wastewater odor was detectable. As 1:200 is the most diluted sample (APHA, 1998), it was not possible to prepare more diluted samples, and the TON resulted was 200 (Table 6). This means that treated wastewater from WWTP of Albacete is not advisable for public use as turf grass irrigation because of the odor nuisance.

Sample volumen diluted to 200 ml (ml)	TON	Sample volumen diluted to 200 ml (ml)	TON
200	1	12.0	17
140	1.4	8.3	24
100	2	5.7	35
70	3	4.0	50
50	4	2.8	70
35	6	2.0	100
25	8	1.4	140
17	12	1.0	200

Table 6. Threshold odor numbers (TON) corresponding to various dilutions (APHA, 1998).

3.2 Soil

Table 7 shows that two months after stopping irrigation, no differences in pH were observed in soils between treatments or depths (0-25 cm, 25-45 cm). Organic matter increased at 0-25 cm in depth with respect to the original soil, while at the end of the study organic matter content at 25-45 cm in depth was lower in both types of soil. No important differences were observed in nitrogen content between treatments or depths, but a slight decrease at the end of the study was observed. C:N ratio increased for the two treatments in the 0-25 layer, but in the deeper layer it was the same as in the original soil. In general, nitric nitrogen in soils varies a lot, and in our case, at the end of the study was lower in all stratums than original soil. There were no differences in the 0-25 cm layer for both treatments at the end of the study, and at 25-45 cm in depth the value was lower than in the upper layer. At this depth, the nitric nitrogen level was higher in wastewater-irrigated soil than in the control soil. Phosphorus content in soil decreased at the end of the study respect to original soil. Two months after ceasing irrigation, wastewater-irrigated soil had higher phosphorus content than the control soil for the same depth and, to more depth, less phosphorus content in soil for the same treatment.

PARAMETERS	UNITS	November 2008			
		Control Plot	Wastewater Irrigated Plot	Control Plot	Wastewater Irrigated Plot
		0-25 cm		25-45 cm	
Sand	%	28	23	43	40
Silt	%	28	28	23	20
Clay	%	45	50	35	40
pH		8.57	8.91	8.45	8.51
E.C	mmhos cm^{-1}	0.31	0.43	0.49	0.70
Chloride	(mg gypsum) (100 g soil)$^{-1}$	44	59	68	133
Sulphate	(mg gypsum) (100 g soil)$^{-1}$	37	115	163	188
Organic matter	%	3.6	3.7	1.8	2.0
Total N	%	0.15	0.15	0.17	0.13
C:N Ratio		14	14	7	9
Nitric N	mg kg^{-1}	3	3	1	2
Asimilable P	mg kg^{-1}	35	41	19	26
Assimilable K	meq 100g^{-1}	2.05	2.47	1.02	0.98
Assimilable Na	meq 100g^{-1}	0.59	2.04	0.64	2.20
Assimilable Ca	meq 100g^{-1}	33.25	30.88	27.31	24.31
Assimilable Mg	meq 100g^{-1}	6.09	6.40	4.46	3.95
Assimilable B	mg kg^{-1}	2.40	5.60	2.39	2.40
K:Mg Ratio		0.34	0.39	0.23	0.25
Ca:Mg Ratio		5	5	6	6

Table 7. Chemical characteristics of the soil two months after ceasing irrigation at 0-25 cm and 25-45 cm in the control plot and wastewater-irrigated plot.

No changes in potassium soil content were observed throughout the study period but at the end of the study the deeper layer had less potassium. Magnesium increased in both types of treatment at 0-25 cm, and there were no differences between them for the same depth. In any case, magnesium content was higher at greater depth. A slight increase in calcium content at the end of the test was observed in the 0-25 cm layer with respect to the original soil. Hardly any differences between treatments were recorded, although calcium in the control soil was slightly higher.

No differences in K:Mg between treatments were observed for the same depth. This ratio remained stable in the 0-25 cm stratum but in at greater depth the value was lower. Finally, the Ca:Mg ratio remained constant throughout the study period but a slight increase at the end of the study at 25-45 cm in depth was observed. No differences between treatments were seen.

Usually, treated wastewater is more saline than tap water, and therefore, when it is reused in irrigation, more salinity problems can occur (Beltrao et al., 2003). Electrical conductivity decreased at the end of the study and was higher in the lower layer. In our case, the sodium content in irrigated water (170 mg l^{-1} in wastewater and 13.6 mg l^{-1} in control water) caused a noticeable increase of this salt in wastewater-irrigated soil at both depths studied. This must be kept in mind because sodium not only affects the soil structure, but may also have a toxic effect on plants. Previous research in this area showed E. C. stability (Mañas, 2006). In this study, an increase in sodium did not increase electrical conductivity, although several studies have shown that the higher the salt concentration is, the higher electrical conductivity is as well (Glober, 1996; Urbano, 2001; Mohammad and Mazahreh, 2003). On the other hand, sulphate content in the upper layer of the control soil decreased at the end of the study (37 (mg gypsum) (100 g soil)$^{-1}$) but not in the rest of layers, so wastewater-irrigated soil had a higher sulphate level than control soil and in both cases sulphate concentrations were higher in deeper layers. Although there were no significant differences between sulphate content in wastewater and drinking water, higher sulphate was observed in the first 25 cm of soil irrigated with wastewater and also in the deeper layer for the two treatments at the end of the study. The sulphate ion causes no particular harmful effects on soils or plants; however, it contributes to increased salinity in the soil solution (Glober, 1996).

Otherwise, the most common phytotoxic ions that may be present in municipal sewage and treated effluents in concentrations such as to cause toxicity are boron, chloride and sodium (Pescod, 1992; Quian, 2008). The sodium content in the control soil was lower (0.59 meq $100g^{-1}$ at 0-25 cm and 0.64 meq $100g^{-1}$ at 25-45 cm) than in wastewater-irrigated soil (2.04 meq $100g^{-1}$ at 0-25 cm and 2.20 meq $100g^{-1}$ at 25-45 cm) in both layers analyzed. Chloride increased at the end of the study in both cases, and for the same treatment there was more chloride at greater depth. There is a scarcity of specific research on the function of boron in turf grasses. Plants differ in their boron requirements, with grasses generally having a much lower demand than dicotyledonous plants (Hull, 2002). In this study, boron content in the wastewater-irrigated plot was higher in the upper soil layer at the end of the study. Assimilable B content increased in the upper soil layer irrigated with wastewater (5.60 mg kg^{-1}). This fact could be the cause of phytotoxicity problems in several crops. In soils organic matter, Fe^{3+} and Al^{3+} oxides retain boron because of complex formation. This causes a very strong energetic fixation which makes it difficult for the plant to absorb boron. Maximum boron adsorption in soil occurs with $Al(OH)_3$ at a pH of 8-9 (Urbano, 2001). At the end of our study, assimilable boron is more than double (5.60 mg kg^{-1}) in the wastewater-irrigated plot than in the control plot (2.40 mg kg^{-1}). Hull (2002) cites some experiences on the effects of managing turf using boron-contaminated irrigation water. In dry climates, boron from irrigation water can accumulate to concentrations of 10 ppm or greater in the soil. However, their research indicated that if turf is growing rapidly, it will dilute boron sufficiently that it will not reach toxic levels in plant tissues. In our case, the growth rate in the wastewater-irrigated plot was very fast. Therefore, since boron content in the irrigation water and soil after two years is not high we can deduce that the amount of boron in turf leaves did not increase to near toxic levels. As we explained previously, boron content in wastewater it is not a problem even for crops sensitive to boron. Hence, we do not consider boron an impediment for turf grass wastewater irrigation.

3.3 Turf grass
3.3.1 Height and phytomass
Grass height was recorded after the first mowing. Measurements show that the height of the wastewater-irrigated crop was always significantly higher than the control crop (Figure 6). The speed of growth after each mowing was higher for wastewater-irrigated grass as well.

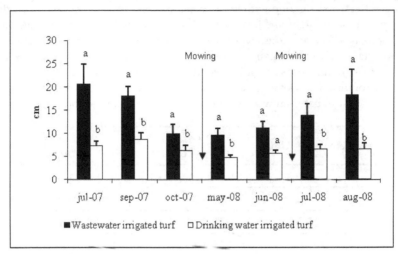

Fig. 6. Turf grass height (cm) throughout the study period.

During the winter season (October 2007-May 2008) the crop was not watered because grass growth ceased, resulting in lower height in both treatments.
In July and August, 2008 (after the second mowing), the wastewater-irrigated grass continued growing until the end of the test, whereas the control grass maintained its height with no further growth reported.

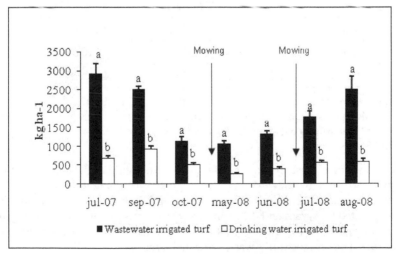

Fig. 7. Phytomass yield (kg ha⁻¹) throughout the study period.

The phytomass yield graph shows the same shape as the height graph (Figure 7). Growth in the wastewater-irrigated crop was always significantly higher than in the control crop.

Studies have shown that continued use of irrigation water with electrical conductivity exceeding 0.75 dS m^{-1} or total dissolved salts greater than 480 mg kg^{-1} may reduce the growth and quality of turf grasses (Camberato, 2001). This is not the case in our study. Turf responded positively to irrigation with wastewater and reached greater height in addition to generating more biomass than turf irrigated with drinking water. Environmental conditions could substantially affect turf grass salt tolerance (Suplick-Ploense et al, 2002) and, in this case, the adequate nutrient content in irrigation wastewater, as shown below, and the selection of a resistant turf grass species avoided this problem.

Regarding macronutrients in leaves, no significant differences in P, K and Mg foliar content were observed (Table 8) between wastewater-irrigated grass and the control.

Parameter	Units	Treated Wastewater	Drinking Water
		Average±sd	Average±sd
N	%	2.18±0.10 a	1.69±0.06 b
P	%	0.22±0.01 a	0.23±0.01 a
K	%	1.83±0.04 a	1.78±0.06 a
Ca	%	0.66±0.02 a	0.82±0.03 b
Mg	%	0.33±0.01 a	0.34±0.01 a
Na	mg kg^{-1}	2291.50±66.78 a	280.47±9.08 b
Fe	mg kg^{-1}	57.63±2.65 a	117.43±10.45 b
Mn	mg kg^{-1}	52.65±1.85 a	56.15±0.26 a
Zn	mg kg^{-1}	17.70±6.20 a	17.17±0.33 a
Cu	mg kg^{-1}	4.93±0.12 a	4.70±0.08 a
Al	mg kg^{-1}	53.53±3.89 a	146.63±17.67 b

Table 8. Chemical composition on Nitrogen, Phosphorus, Potassium, Calcium, Magnesium, Sodium, Iron, Manganese, Zinc, Copper and Aluminium in turf grass (dw). Different letters mean significant differences between Treated Wastewater and Drinking Water at $p<0.05$, according to LSD test. (n=4).

Nevertheless, one of the most important differences with regard to the control turf grass was that turf grass irrigated with wastewater had greater N content (2.18% in wastewater irrigated turf grass and 1.69% in control plants) and less Ca in plant tissue (0.66% in wastewater irrigated turf grass and 0.82% in control turf grass). This higher nitrogen content in wastewater irrigated turf is directly linked with the higher phytomass yield obtained in this kind of turf. The present study was performed over two years, but other studies show that a longer period of wastewater application (10 years) resulted in lower biomass production which nonetheless remained higher than that of the control plants (Mohammad et al., 2007). Thus, periodic monitoring of soil and plant quality parameters would be recommended to ensure successful, safe, long-term wastewater irrigation.

With respect to micronutrients, it is important to emphasize the behavior of Na because foliar tissue from the wastewater-irrigated plot is much higher (2291.5 mg kg⁻¹) than the control (280.5 mg kg⁻¹). Thus, Na in wastewater-irrigated turf grass leaves was 8 times higher with respect to the control (717% higher). This may be an effect of sprinkler irrigation, as sodium and chloride frequently accumulate by direct adsorption through leaves that are moistened (Quian, 2008).

Fe and Al foliar content was significant lower in wastewater-irrigated grass than in the control. This could be because Fe and Al are fixed by organic matter in the soil, and the plant cannot take them up. Finally, no significant differences were observed for Mn, Zn and Cu.

If the plants were grown for raw consumption, heavy metal contamination of urban agricultural fields under long-term application of wastewater could be problematic, and there are many studies on this topic (Mañas, 2006; Castro, 2007; Agbenin et al., 2009). However, this is not the case in the present study.

Regarding microbiological parameters, the main factors that affect pathogen survival in the environment are humidity, soil content, temperature, pH, sunlight (ultraviolet radiation), foliage and competition with native flora and fauna; pathogen inactivation is much more rapid in hot, sunny weather than in cool, cloudy and rainy conditions and low temperatures prolong pathogen survival (WHO, 2006).

Sunlight and temperature are parameters with a high influence on the dynamics of microbes, and in our study we can see that the slopes of the mean average temperature and sunlight (Figure 3) fit into a multiplicative model (Equation 2). Y= mean average temperature; X= Sunlight:

$$Ln\ (Y) = -1.71244 + 1.46425 * Ln(X) \tag{2}$$

$$R^2 = 90.52\% \text{ and } P\text{-value} < 0.001$$

Since sunlight has a strong correlation with temperature, we can select either of the two variables to study the evolution of pathogens. Hence, Figures 8, 9, 10 and 11 show the dynamics of microorganisms related to sunlight in a regression analysis in which independent variable is "days after planting" and cfu count is considered dependent variable. It is known that high temperatures lead to rapid mortality and low temperatures lead to prolonged survival, while freezing temperatures can also cause pathogen mortality. Direct sunlight leads to rapid pathogen inactivation through desiccation and exposure to ultraviolet radiation (WHO, 2006).

In the present study, sunlight decreased from July 2007 to January 2008 and increased from then until the end of August, 2008 (Figure 3). *Salmonella sp* was not detected in leaves at any time. Total coliforms and faecal streptococci in turf grass leaves (Figures 8 and 10) increased with both types of irrigation (wastewater and control). Faecal coliforms (Figure 9) also increased but the trend was more stable than the trend for the microbes mentioned above.

The unchanging increase in microorganisms could be explained because in spring-summer time, constant irrigation of turf with both types of water led to microbial growth despite the

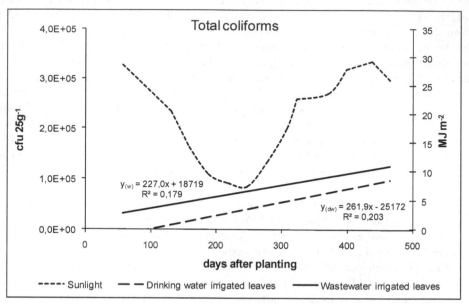

Fig. 8. Dynamics of total coliforms in turf grass leaves irrigated with drinking water versus wastewater and sunlight during the study period.

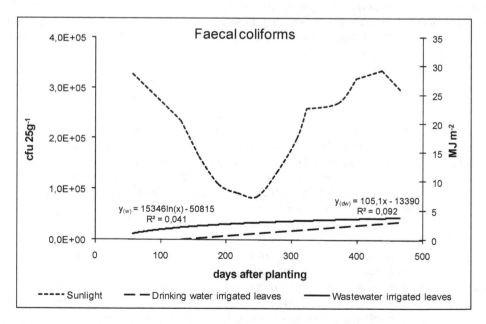

Fig. 9. Dynamics of faecal coliforms in turf grass leaves irrigated with drinking water versus wastewater and sunlight during study period.

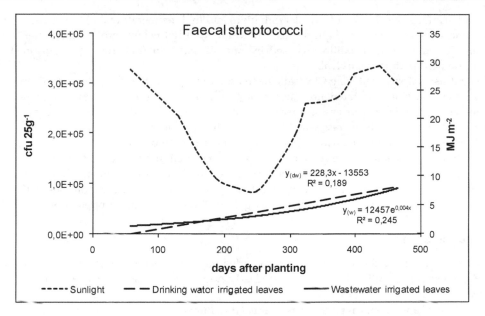

Fig. 10. Dynamics of faecal streptococci in turf grass leaves irrigated with drinking water versus wastewater and sunlight during study period

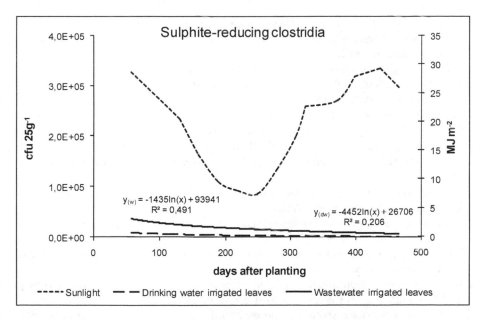

Fig. 11. Dynamics of sulphite-reducing Clostridia in turf grass leaves irrigated with drinking water versus wastewater and sunlight during study period.

negative effects of sunlight (ultraviolet radiation) and high temperature on pathogens. In the autumn-winter seasons, sunlight decreased but low (but not below zero) temperatures and humid environmental conditions caused by rainfall that year (not from irrigation in this case) favored pathogen survival.

By contrast, sulphite-reducing Clostridia decreased throughout the study (Figure 11). In our case, constant irrigation maintains microbiological population on the leaves. Although the dynamics of microorganisms showed parallel curves, the slope of microbes from wastewater-irrigated turf grass was higher than the control in the case of total and faecal coliforms and sulphite-reducing Clostridia (Figures 8, 9 and 11). In contrast, Figure 10 showed no significant difference in faecal steptococci content in the two treatments.

We can deduce that the dynamics of total and faecal coliforms and sulphite-reducing Clostridia were the same for both types of irrigation water but the continuous entry of microorganisms in plots with wastewater irrigation allowed for a larger population.

4. Conclusions

To conclude our study we have find some advantages and some disadvantages in using wastewater for irrigation turf grass. It is clearly noticeable that some of those advantages are:

- Physical and chemical parameters evaluated in our study showed that the use of treated wastewater for irrigate turf grass was useful to get improve in its growth, especially in terms of height and phytomass.
- No negative effects with respect to changes in soil pH, electrical conductivity or salinity occurred.
- Treated wastewater from the Albacete wastewater treatment plant can be a source of fertilizer since it has an important contribution of N, P and organic matter and can save farmers money on fertilizer.
- In foliar tissue, *Salmonella sp.* was not detected in any case and sulphite-reducing clostridia content did not increase throughout the study period.

However, it is necessary to be aware that several crops could suffer from the negative effects of wastewater irrigation due to the sodium and boron content and should keep this in mind that the benefits of irrigation with treated urban wastewater including addition of plant nutrients to turf grass and conservation of valuable freshwater resources.

In contrast, some risks for human health arise from microbiological aspects also evaluated in this study:

- Helminth egg content in wastewater irrigation results were 6 eggs 10 l-1 and exceeded the limit recommended by the WHO and Spanish legislation.
- One hundred percent of testers agreed that wastewater odor was perceptible at a 200 ml dilution.
- Coliforms (total and faecal) and faecal streptococci have a tendency to increase throughout the study.

In spite of the commented advantages and according to the odor test and microbiogical parameters results it is not advisable to use wastewater irrigation from the wastewater treatment plant of Albacete since it could be harmful in a public garden. It is essential to disinfect this wastewater and remove the odor with appropriate advanced treatments at the end of the process before using it.

5. Acknowledgments

The authors are grateful to the Town Hall of Albacete for financial support through the project: "Estudio sobre la calidad de las aguas de abastecimiento de la provincia y la aptitud agronómica de aguas y lodos generados en la E.D.A.R. de Albacete."
Thanks to Stefanie Kroll for reviewing the English text.

6. References

Agbenin, J.O. ; Danko, M. & Welp, G. (2009). Soil and vegetable compositional relatioships of eight potentially toxic metals in urban garden fields from northern Nigeria. *J. Sci. Food Agric.*, 89, pp: 49-54.

Allison, L.E. (1965). Organic Carbon, in Methods of Soil Analysis, ed. by Black CA, Evans DD, White JL, Ensminger LE and Clark FE, Madison, Wisconsin: *American Society of Agronomy*, pp: 1367-1378.

Angin, I. ; Vahap, A. & Turan, M. (2005). Effects of long-term wastewater irrigation on soil properties. *Journal of Sustainable Agriculture*, 26(3), pp: 31-42.

APHA. (1998). Standard Methods for the Examination of Water and Wastewater, 20th edition, 1998, *American Public Health Association, Washington, DC.*

Ayers, R.S. & Westcot, D.W. (1987). Water quality for agriculture. *Irrigation and Drainage Paper 29 Rev. 1.* FAO, Rome. 174 pp.

Barton, L. ; Schipper, L.A. ; Barkle, G.F. ; McLeod, M. ; Speir, T.W. ; Taylor, M.D. ; McGill, A.C. ; van Schaik, A.P. ; Fitzgerald, N.B. & Pandey, S.P. (2005). Land application of domestic effluent onto four soil types: plant uptake and nutrient leaching. *J. Environ. Qual.*, 34, 635-643.

Beltrao, J.;, Costa, M.;, Rosado, V.,; Gamito, P.;, Santos, R. &;, Khaydarova, V. (2003). New techniques to control salinity-wastewater reuse interactions in golf courses of th mediterranean regions. *Geophysical Research Abstracts*, 5, 14168.

Bo, Y.; Wei, J.; Fen, L.; Zhenlong, G. & Anxi, J. (2001). Preparation of activated carbons from sewage sludge and removing hydrogen sulfide. *Advanced Materials Research*, (Vols. 183 - 185), pp: 1332-1335.

Bower, C. & Wilconx, L. (1965). Soluble salts in Methods of Soil Analysis, ed. by Black C.A.;, Evans D.D.,; White J.L.,; Ensminger L.E. and& Clark F.E. ;, Madison, Wisconsin: *American Society of Agronomy*, pp: 914-926.

Bremmer, J.M. (1965). Total nitrogen, Soluble salts, in Methods of Soil Analysis, ed. by Black C.A.; Evans D.D.; White J.L.; Ensminger L.E. & Clark F.E. ; Madison, Black, C.A,, Evans, D.D., White, J.L., Ensminger, L.E., Clark, F.E. Madison, Wisconsin: *American Society of Agronomy*, pp: 1149-1178.

Camberato, J.J. (2001). Turf Irrigation Water Quality Slide Set,. Available from http://virtual.clemson.edu/groups/psapublishing/ppt_files/Turfwtr_files/v3_do cument

Campos, C. (2008). New perspectives on microbiological water control for wastewater reuse. *Desalination*, (218) pp: 34-42.

Capelli, L.; Sironi, S.; Del Rosso, R. &; Céntola, P. (2009) Predicting odour emissions from wastewater treatment plants by means of odour emission factors. *Water Research*, Vol. 43, pp: 1977-1985.

Castro, E. (2007). Aptitud para uso agrícola del agua residual depurada y lodos generados en una estación depuradora de aguas residuales de fangos activos. Tesis Doctoral. *Universidad de Castilla-La Mancha. Escuela Técnica Superior de Ingenieros Agrónomos de Albacete. Departamento de Producción Vegetal y Tecnología Agraria.*

Castro, E.; Mañas, M.P. & de las Heras, J. (2011). Effects of wastewater irrigation on soil properties. *Water Science & Technology* Vol. 63.8; pp: 1679-1688.

Cheng, X.; Wodarczyk, M.; Lendzinski, R.; Peterkin, E. &; Burlingame, G.A. (2009). Control of DMSO in wastewater to prevent DMS nuisance odors. *Water Research*, Vol. 43, pp: 2989-2998.

Chizuru, A.; Mushtaq, A.M. & ; Hiroyuki, M. (2005). Water and wastewater reuse. An environmentally sound approach for sustainable urban water management. In: *United Nations Environment Programme.*

Davis, C. (2011). *E. coli 0157:H7 Facts.* Available from http://www.medicinenet.com/e_coli__0157h7/article.htm.

De las Heras, J.; Mañas, P. &; Labrador, J. (2005). Effects of several applications of digested sewage sludge on soil and plants. *Journal of Environmental Science and Health*, A40, pp: 437–451.

DOH. Washington State Department of Health. (2007). Coliform bacteria and drinking water. *Division of Environmental Health Office of Drinking Water.* pp: 331-181.

Doorenbos, J. &; Pruitt, W. (1984). Las necesidades de agua de los cultivos. Paper 24. *FAO. Roma.*

Gan, J. S.;, Bondarenko, F.;, Ernst, W.,; Yang, S.,; Ries, B. &, Sedlak, D.L. (2006). Leaching of N-Nitrosodimethylamine (NDMA) in turf grass soils during wastewater irrigation. *J. Environ. Qual.*, Vol. 35, pp: 277-284.

Glober, C.R. (1996). Irrigation water classification systems. Guide A-116. *Cooperative Extension Service. College of Agriculture and Home Economics.*

Hull, R.J. (2002). Recent research offers clues to boron's purpose,. Available from http://www.tufrgrasstrends.com.

Jiménez, B. (2007). Helminths (worms) eggs control in wastewater and sludge. *International Symposium on New Directions in Urban Water Management.* 12-14 September 2007, UNESCO Paris.

Lawrence, C.C. &; Tan, N.C. (1990). Odour generation potential of wastewaters. *Water Research*, 24 (12), pp: 1453-1458.

Lockett, A.M. (2008). Impact of reuse water on golf course oil and turf grass parameters monitored over a 4.5 year period. *HortScience*, Vol. 43, pp: 1942-2274.

Mañas, P. (2006). Aptitud agronómica del agua residual depurada y lodos generados en una Estación Depuradora de Aguas Residuales de lechos bacterianos. Tesis Doctoral, 2006, *Universidad de Castilla-La Mancha. Escuela Técnica Superior de*

Ingenieros Agrónomos de Albacete. Departamento de Producción Vegetal y Tecnología Agraria.

Mañas, P.; Castro, E.; Vila, P. &; De las Heras, J. (2010). Use of waste materials as nursery growing media for Pinus halepensis production. *Eur. J. Forest Res.* Vol 129, pp: 521-530. DOI 10.1007/s10342-009-0349-4.

McKenzie, C. (2005). Wastewater reuse conserves water and protects waterways. *On Tap*, pp: 46-51.

Menzel, C.M. & Broomhall, P. (2005). Recycled water: Good for parks, gardens and the environment. *Proceedings of the International Symposium on Harnessing the Potential of Horticulture in the Asian-Pacific Region.* Acta Horticulturae, (694), 191-196.

Mohammad, M.J. & Mazahreh, N. (2003). Changes in Soil Fertility Parameters in Response to Irrigation of Forage Crops with Secondary Treated Wastewater. *Communications In Soil Science And Plant Analysls*, 34, (9-10), 1281-1294.

Mohammad, M.J.,; Hinnawi, S. &, Rousan, L. (2007). Long term effect of wastewater irrigation of forage crops on soil and plant quality parameters. *Desalination*, Vol. 215, pp: 143-152.

Olsen, R.L. & Dean, L. (1965). Phosphorus, in Methods of Soil Analysis, ed. by Black C.A.; Evans D.D.; White J.L.; Ensminger L.F. & Clark F.E.; Madison, Wisconsin: *American Society of Agronomy*, 1035-1049.

Peech, N. (1965). Hydrogen ion activity, in Methods of Soil Analysis, ed. by Black C.A.; Evans D.D.; White J.L.; Ensminger L.E. & Clark F.E.; Madison, Wisconsin: *American Society of Agronomy*,914-926.

Pescod, M.B. (1992). Wastewater treatment and use in agriculture. *FAO. Irrigation and drainage.* Paper 47.

Quéméner, J. (1985). L'interpretacion des analyses. *Cultivar*, Vol. 22, pp: 107-117.

Quian, Y.L. & Mecham, B. (2005). Long-term effects of recycled wastewater irrigation on soil chemical properties on golf course fairways. *Agronomy Journal*, Vol. 97, pp: 717-721.

Quian, Y.L. (2008). Recycled wastewater instigates different responses in turfgrass, trees and soils, Available from. http://www.tufrgrasstrends.com.

Real Decreto 1620/2007, de 7 de diciembre, por el que se establece el régimen jurídico de la reutilización de las aguas depuradas. BOE núm. 294.

S.A.F. (Servicio de Asesoramiento a la Fertilización). (2005). *ITAP*, Available from http://www.itap.es/

Salgot, M.; Huertas, E.; Weber, S.; Dott, W. &; Hollender, J. (2006). Wastewater reuse and risk: definition of key objetives. *Desalination*, Vol. 187, pp: 29-40.

Shuval, H.I.; Wax, I.; Yekutiel, P. & Fattal, B. (1989). Transmission of enteric disease associated with wastewater irrigation: a prospective epidemiological study. *Am. J. Public Health*, Vol. 79(7), pp: 850-852.

Suplick-Ploense, M. R.; Quian, Y. L. & Read, J.C. (2002). Relative NaCl tolerance of Kentucky Bluegrass, Texas Bluegrass, and their hybrids. *Crop Science*, Vol. 42, pp: 2025-2030.

Urbano, P. (2001). Tratado de Fitotecnia General. 2ª edición. *Editorial Mundiprensa.*

USEPA. (1992). Guidelines for Water Reuse. EPA 625/R-92/004, *Environmental Protection Agency*. US Washington DC.

WHO. (1989). Health guidelines for the use of wastewater in agriculture and aquaculture. *World Health Organization, Geneva.*

WHO. (2006). Guidelines for the safe use of wastewater excreta and greywater. *World Health Organization.*

Wu, L.; Chen, J.; van Mantgem, P. & Ali Harivandi, M. (1996). Regenerant wastewater irrigation and ion uptake in five turfgrass species. *Journal of plant nutrition*, Vol. 19 (12), pp: 1511-1530.

Yañez, J. (1989). Análisis de suelos y su interpretación. *Horticultura*, Vol. 49, pp: 79-89.

12

Occurrence and Survival of Pathogenic Microorganisms in Irrigation Water

Nohelia Castro-del Campo,
Célida Martínez-Rodríguez and Cristóbal Chaidez
Centro de Investigación en Alimentación y Desarrollo, A.C. Culiacán, Sinaloa,
México

1. Introduction

The consumption per capita of fresh vegetables has increased in the last years in the USA and other countries, which has contributed to the increase of gastroenteritis outbreaks attributed to contaminated fruits and vegetables. Fresh produce can incorporate pathogenic microorganisms thru the process of irrigation, harvesting, postharvest processing and distribution. Most microorganisms use irrigation water and/or soil as a vehicle of transport (Beuchat, 1995; Bhagwat, 2003). Untreated water is most likely to transmit several microorganisms, which may include pathogenic strains of *Escherichia coli*, *Salmonella*, *Listeria*, protozoa and viruses (Díaz et al., 1999). Studies in different countries indicate that the use of untreated water for irrigation of vegetables is the practice most related to fresh produce safety issues (Monge et al., 1996; Díaz et al., 1999; Tyrrel y Quinton, 2003).

Surface water may pose a risk of contamination if its source is unknown. Water is used for multiple issues in diverse agricultural activities including application of fertilizers, washing and disinfecting produce. Therefore, water has to meet the chemical and microbiological requirements before its use (Siller et al., 2002). The quality of water is based on the amount of indicator microorganisms. The major source of coliform contamination when vegetables are grown is probably the irrigation water (Okafo et al., 2003). There are critical factors that need to be monitored to ensure safe water supply. All water sources must be examined periodically for microbiological determination, the results must be recorded and existing problems corrected, for example bathing and grazing animals nearby water resources which should be prohibited to prevent fecal contamination and reduce risks to human health from consuming contaminated fresh produce.

Frequently, contamination is associated with the application of irrigation water and the type of crop. Studies have proved that flooding irrigation represents the greatest possibility of contamination if it's used on produce having direct contact with the soil while the sprinkler irrigation technique provides a rapid means to contaminate the product if the water is contaminated; On the other hand, the drip irrigation technique has represented the lowest risk of contamination of produce (Siller-Cepeda et al., 2009). Lettuce, radishes, carrots are vegetables most likely to become contaminated due to their direct contact with soil and water which can possibly contain bacteria (Okafo et al., 2003). It is well recognized that fecal indicator bacteria may be transported and be a source of contamination of water sources (Tyrrel and Quinton, 2003).

Fecal and total coliforms are considered the index parameter of pathogenic microorganisms, but recently this role has been questioned because of its ability to grow in various water sources (Gleeson, 1997; Banning et al., 2002).

Since its first isolation and description in 1982, E. coli O157:H7 has been associated with numerous outbreaks in the US and some other countries in the world (Bitton, 1994). From the foodborne outbreaks occurred in Japan, the 1996 radish outbreak is among the most deadly with thousands of people ill and 11 school children deaths. Scotland also suffers from an outbreak due to the consumption of undercooked meat that caused the deaths of 16 elderly (Loaharanu, 2001).

Listeria monocytogenes is a Gram positive organism found in the environment and cause disease in animals and humans. The host susceptibility is important for the development of the disease having a primary target pregnant woman, immunocompromised persons, elderly and newborns. One of the most important properties is the intrinsic resistance to the environment and its growth at broad ranges of pH (4.5 – 9), temperature of 0 to 45°C and Aw (0.92) (Copes et al., 2000). This bacterium can survive for long periods in soil and crops; it does not lose its virulence during the period on the soil or the environment becoming a health hazard if ingested thru contaminated fresh produce. The route of transmission of Listeria is by the ingestion of contaminated food and water (Al-Ghazali and Al-Azawi, 1990).

Salmonella include more than 2500 pathogenic serotypes. It is ubiquitous in the environment and is commonly found in water, food and other materials. This pathogen colonizes the intestinal tract of humans, animals, birds and insects and is transmitted to the irrigation water, drinking water, mild and other raw foods by fecal contamination (Abushelabi et al., 2003). It is also commonly found in reusable water and contaminate by runoff of rivers, discharges into the sea and other sources of agricultural water (Lemarchand and Philippe, 2002).

Microbiological contamination of pathogenic bacteria in vegetables becomes more important when considering that survival time may be prolonged for weeks or months, particularly when they are protected from desiccation and direct sunlight, as in lettuce, cabbage and radish. Studies have shown that pathogens inoculated into farmland or irrigation water can survive for up to two months, a period sufficient to reach the consumer (Monge et al., 1996).

Survival is defined as the ability to maintain viability with adverse circumstances. Bacteria in water respond to different physical and chemical variables including low or high concentration of dissolved oxygen, redox potential and pH (Roszak and Colwell, 1989). When bacteria get in contact with the surface of vegetables can initiate a process of adhesion and subsequent colonization under favorable conditions. Some cells produce extracellular polymers that lead to the formation of biofilms. This process is usually accompanied by an increased tolerance to the effect of drying and the lethal action of ultraviolet light. Concurrently microorganisms protect themselves staying at sites such as cracks, depressions, stomata, lenticels and trichomes that are part of the structure of vegetation (Fernández, 2001).

The existing national standards establish the limits of microbial contamination only for wastewater. There are no rules governing the use of irrigation water at the national or international level. The amount of microorganisms varies considerably in agricultural water and wastewater, this type of water is ruled by the NOM-003-ECOL-1997 that establishes the maximum contaminant limits for treated wastewater to be reused. The FAO/WHO 1989, set as the maximum level of 10^3 fecal coliforms/100 mL (Monge et al., 1996).

One of the problems facing growers of the Valley of Culiacan is the shortage of water due to lack of rain, the indiscriminate use of the agricultural sector and the growing population. The valley of Culiacan is located within 010 irrigation districts consisting of 8 modules of irrigation, of importance to this study are II-1, II-2 and IV-1, which supply water to the main producers of agricultural fresh produce in our state (Table 1).

Fresh Produce Crops	Irrigation Districts		
	Module II-2	Module III-3	Module IV-1
Tomato	✓	✓	✓
Cucumber	✓	✓	✓
Bell pepper	✓	✓	✓
Eggplant	✓	✓	✓
Squash	✓	✓	
Scallion	✓	✓	
Swiss chard	✓		
Onion			✓
Green beans	✓		✓

Table 1. Fresh produce crops irrigated with superficial water through the irrigation modules in the Culiacan Valley.

There are no studies on the presence and survival of these bacteria in water used to irrigate vegetables in the valley of Culiacan. Because of this reason it was proposed to determine the incidence of *Escherichia coli*, *Salmonella* spp and *Listeria* spp in canal water for agricultural use and to determine its survival exposed to different physicochemical parameters in laboratory conditions.

2. Methods

The study was divided in *Field Study* and *Lab Study*:

2.1 Field study

It was performed a correlational descriptive study to identify bacteria and their survival by using a three factor completely randomized repeated measurements. Seventy samples of water from canals in the Culiacan Valley where taken in two stages, the first one from February to May 2003 (fresh produce production ending, rainy season) and the second one from November 2003 to February 2004 (fresh produce production pick, drought season). Sampling points were determined randomly according to the area of greatest fresh produce production in the Culiacan Valley, Sinaloa, specifically in the irrigation modules II-2, II-3 and IV-1 including crops of tomato, cucumber, squash, eggplant among others (Table 1). Sinaloa state is located at the northwest of Mexico (27° 7′-22° 20′N, 105° 22′-109-109° 30′O) (Figure 1) with a population of 2,767,761 people (INEGI, 2011).

Collection of samples

The samples of water were collected in sterile 1-L bottles, properly labeled and sealed. Transportation to the laboratory was conducted under refrigeration (4-6°C) for the microbiological analysis. The study analyzed a total of 70 samples, where the parameters turbidity, temperature and pH were measured in the field.

Microbiological analysis

Fecal coliforms were determined by the membrane filtration technique (APHA, 2001). One and 10 mL of water samples were filtered through a membrane of 0.45 μm and plated on m-FC agar (Difco) and incubated 24 h at 45°C. For the biochemical determination of *E. coli* the API 20E system was used (Biomeriux Vitek, Hazelwood, MO) by placing a positive coliform colony in 5 mL of 0.85% sodium chloride saline solution to obtain a bacterial suspension, which was added in the gallery microtubes and incubated at 37°C during 24 h. Gallery reading was based on the API 20E table. For the isolation of *Salmonella* spp and *Listeria* spp enrichment broths and selective agars were used as describe by APHA (2001).

2.2 Lab study

The second phase of the study consisted in an *in-lab study* performing bacterial survival analysis at different temperature, turbidity and pH. For this, positive samples of *Escherichia coli* and *Salmonella* spp obtained from the *Field Study* and confirmed by API 20E were used. *Listeria* spp was obtained from the State Public Health Laboratory.

Bacterial Regeneration

A colony of the bacterial strains was isolated and transferred separately in 5 mL of soy broth trypticasein (TSB) which was incubated for 24 at 37°C. Once the incubation time elapsed, 1 mL of this bacterial suspension was taken and added in an Erlenmeyer flask containing 50 mL of TSB, being incubated again for 24 h at 37°C. The cultured broth was placed in tubes (Beckman J2-MI) and centrifuged at 10,000 rpm, 4°C for 11 min. The precipitate obtained was re-suspended twice with 15 mL of phosphate buffer (KH_2PO_4 pH 7.2) stirring vigorously with a vortex mixer (Scientific Products) to obtain a homogeneous solution. This process was repeated twice (Abbaszadegan et al., 1997). Finally, the pellet obtained was suspended in 300 mL of buffer reaching a concentration of 1×10^8 colony forming units per milliliter (CFU/mL).

Survival Evaluation

Survival evaluation followed the methodology described by Johnson et al. (1997), with some modifications described below. A liter of canal water was inoculated with 30 mL of the three regenerated bacteria reaching concentrations of 1×10^7 CFU/mL and different temperature (15, 35, 40°C) and turbidity (2, 20, 50 Nephelometric Units, NTU) were adjusted. These mixtures were exposed to the environment with constant stirring for 72 h. Aliquots were taken at different exposure times (0, 3, 12, 24, and 48). Dilutions were performed and 0.1 mL was stricken onto m-FC agar (Difco) for the identification of *E. coli*, Hektoen agar (Difco) for *Salmonella* and Palcam agar (Difco) for *Listeria* and finally incubated at 35 ± 0.5°C for 24 h. Each treatment was performed in duplicate.

2.3 Statistical analysis

The analysis for the quantification of coliforms was average in the three irrigation modules, while for the survival analysis a three factors design completely randomized with repeated measures over time was performed with MINITAB program with a significance level of $p > 0.05$.

3. Results

3.1 Field study

During the *first stage* there was a low rainfall contribution. The highest fecal coliforms average was detected in the irrigation module II-2 with more than 40000 CFU/100 mL (Table 2). The temperature remained constant during the sampling period ranging between 33 and 35°C, behaving similarly than the pH with a value of 7.1 while total chlorine was registered with an average of 0.9 ppm. The difference was observed in turbidity registering levels higher than 50 NTU (Table 2).

Parameter	Module II-2	Module III-3	Module IV-1
Fecal coliforms (CFU/100 mL)	40000	23000	28000
Temperature (°C)	35	33.5	34.5
pH	7.13	7.16	7.13
Turbidity (NTU)	48.43	56.5	17.08
Total chlorine (mg/L)	0.11	0.08	0.08

Table 2. Microbiological and Physicochemical quality of water during the first stage February-May, 2003.

During the *second stage* there was a rainfall contribution and it is called the rainy season. In this *stage* of sampling it was observed fewer fecal coliforms with respect to the first stage, presenting the highest numbers the module III-3 with more than 7500 CFU/100 mL, similar levels to those of the IV-1 module with 7000 CFU/100 mL. The temperature was kept within a range not exceeding 22°C and the presence of total chlorine less than 0.2 ppm (Table 3).The temperature was kept within a range no greater than 21.2 °C and the presence of total chlorine was less than 0.2 parts per million.

Parameters	Module II-2	Module III-3	Module IV-1
Fecal coliforms (CFU/100 mL)	4000	7500	7000
Temperature (°C)	21.2	20.3	20.6
pH	7.08	7.5	7.2
Turbidity (NTU)	30	48	49
Total chlorine (mg/L)	0.07	0.15	0.13

Table 3. Microbiological and Physicochemical quality of water during the second stage November, 2003 – February, 2004.

In table 4 can be observed that the bacterium found in greater proportions during the *first stage* was *E. coli* in the three irrigation modules, accumulating a total of 36 positive samples out of 40 (90%). The presence of *Salmonella* is noted but to a lesser extent with a total of 8 positive samples (20%). In the other hand the presence of *Listeria* spp was not detected.

Irrigation districts	E. coli	Salmonella spp	Listeria spp
II-2	12	2	0
IV-1	10	2	0
III-3	14	4	0
Total	36	8	0

Table 4. Number of positive samples for E. coli,Salmonella and Listeria during the first stage february-May, 2003.

Results of the presence of pathogenic bacteria during the *second stage* show *Escherichia coli* with a highest incidence of 63.3 % positive samples, while *Salmonella* spp. represented only a 6.6 %. Finally, *Listeria* spp was not detected as in the *first stage*.

Irrigation districts	E. coli	Salmonella spp	Listeria spp
II-2	5	1	0
IV-1	8	1	0
III-3	6	0	0
Total	19	2	0

Table 5. Number of positive samples for E. coli, Salmonella and Listeria during the second stage November, 2003 to February, 2004.

It has been shown that the frequent contamination of soil is due to the application of contaminated water or animal and humans feces. However, microorganisms have strategies that enable them to face the adverse environmental factors and allow pathogens to remain viable in soil for two months or more especially in shaded moist areas (Solomon *et al.*, 2002; Monge *et al.*, 1996). This could explain at least partially fecal contamination during the rainy season in those vegetables whose edible portion is in direct contact with the ground (Cifuentes *et al.*, 1994, Monge *et al.*, 1996). Tyrrel y Quinton (2003) mention that the transport of coliform in water is mediated by the density and turbidity of the water, which means when there are continuous rainfall the volume of water bodies increases, causing a dilution in the total coliforms originally present which coincides with the present investigation founding decreased concentrations in the second stage sampled in different modules. In spite of the low concentrations, epidemiological studies have shown that there is a risk of gastroenteritis, when people are indirect contact with 20 to 35 CFU /100 mL fecal population (Polo et al., 1998). In the present study there were found higher levels of fecal contamination in canal which represent a risk of contamination for vegetables that come into contact with it and furthermore for consumers.

Polo *et al.* (1998) reported absence of *Salmonella* in the presence of fecal coliforms in river water and can be explained by interference of aquatic micro biosphere, moreover the authors mention the possibility of mutation that has bacteria in order to be found in the environment as a viable but not cultivable. Okafo *et al.* (2003) found a relationship between the incidence of pathogenic bacteria of 1 *Salmonella* per 45 total coliforms and 13 fecal coliform in water samples. However, the most conclusive result will always be to identify pathogenic bacteria rather than index organisms. Table 4 shows a higher proportion of

Escherichia coli in general making a total of 19 positive samples out of 30. *Salmonella* spp was isolated in a lower proportion (2 positive samples) while *Listeria* was not detected in any of the samples. *Listeria* compared with the Gram negative bacteria shows a survival decrement in aquatic environments, its characteristics as Gram positive bacteria confers the cell with poor resistance to those conditions (Liao and Shollenberger, 2003). In this sense, Monfort *et al.* (2000) refers to *Listeria* as a bacterium resistant in food, refrigerated temperature and low oxygen conditions, however they also mark its low survival in aquatic environments at different stress conditions.

3.2 Lab study

Escherichia coli survival

The survival profile shown in all treatments was descendent, but different between factors. At temperature of 15°C with a 50 NTU turbidity reaching the maximum time of survival registered 4.3 $\log_{10}$ in contrast at temperature of 35°C with 50 NTU turbidity which reached a concentration of 3.4 $\log_{10}$ at the end of the experiment, the difference is greater at 40°C and at 50 UNT reaching concentration of 2.7 $\log_{10}$. This indicates that the high turbidity and lower temperature increase the survival time for *Escherichia coli* under these conditions.

The temperature factor was analyzed showing 35°C as the best temperature for the *E. coli* survival (Figure 1A). When analyzing this factor using the Tukey test (p<0.05) showed statistical differences. Numbers under the same line are not statistical different.

15   <u>35   40</u>

According to the principal effects graphic the *Escherichia coli* survival order based on turbidity was 50 UNT > 20 UNT > 2 UNT (Figure 1B). Statistical differences were observed when the Tukey test was performed (p<0.05).

<u>50   20</u>   2

Along time bacterial concentration presented a considerably decrement (Figure 1C). Four different contact times where statistical different (p<0.05).

<u>48   24   12   0</u>

E. coli presented a longer survival time when exposed to 50 NTU finding a correlation between turbidity and survival time. In other words, the increased turbidity improved survival. When organic matter is present in aquatic environments there is a protective effect of bacteria on organic matter, against the direct rays of the sun (radiation, UV light), which coincides with the results obtained by Mezriqui *et al.* (1995) who conducted a study with *E. coli* and *S.* Typhimurium in sea water finding that water with higher turbidity contained higher concentrations of bacteria and survived for longer period of time. A similar study in which the survival time was measured at different total suspended solids (TSS) concentrations and different solar radiation, found significant differences in solar radiation with respect to TSS, obtaining that a higher radiation and lower TSS further reducing the survival in stagnant water, finding a relationship between increased bacterial survival and greater amount organic matter with respect to time.

In aquatic environments nutrient limitation is a major stressor for the cells. Gram-negative, particularly *E. coli* and *Salmonella* generally increase resistance to environmental factors at the stationary phase. *Escherichia coli* have the ability to survive in aquatic environments even at low temperatures (Rice and Johnson, 2000).

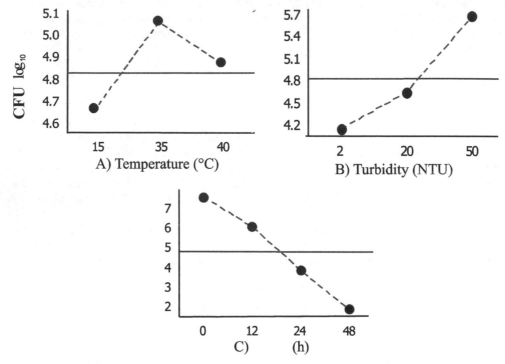

Fig. 1. Principal effects graphic of the three abiotic factors on *Escherichia coli* survival. A) Different levels of the temperature factor, B) Different levels of the turbidity factor, C) Different levels of the time factor.

Salmonella sp Survival

During the 48 h experiment *Salmonella* sp showed a steady decline, but different between the levels of temperature. The temperature of 15°C showed the highest survival over time getting the highest growth at 12 h (6.3 $\log_{10}$) and decreasing its concentration at 24 h to 3.9 $\log_{10}$, to finally registering a concentration of 1.5 $\log_{10}$ at 48 h (Figure 2C).

The survival displayed by *Salmonella* sp was greater at 50 NTU in the course of time (Figure 2A). The initial concentration was 7.4 $\log_{10}$, which was then descending considerably to reach 6.2 $\log_{10}$ after 12h of exposure of abiotic factors. In this sense at 24 h presented a concentration of 4.5 $\log_{10}$ to finally obtain 2.5 $\log_{10}$ at the end of the experiment (48h).The turbidity of 2 and 20 showed a lower bacterial concentration in the course of time detecting levels up to1 $\log_{10}$ after 48 h.

Salmonella sp survival was high when combining 15°C and 50 NTU. The condition of survival of bacteria over time was enhanced by the effect of high turbidity and low temperature.

ANOVA analysis shows that most variability between treatments was due to the effect of turbidity, temperature and time, which were statistically significant (p <0.05).
According to the main effects plot bacterial survival between different turbidities showed differences and by Tukey test (p<0.05) differences were significant (Figure 2A).

<u>2   20</u>   50

For temperature significantly differences (Figure 2B) were detected by the Tukey test (p<0.05).

15   <u>35   40</u>

Regarding the time factor represented a cause of decreasing variability in the course of the experiment on *Salmonella* sp survival (Figure 2C).

<u>48</u>   <u>24</u>   <u>12</u>   0

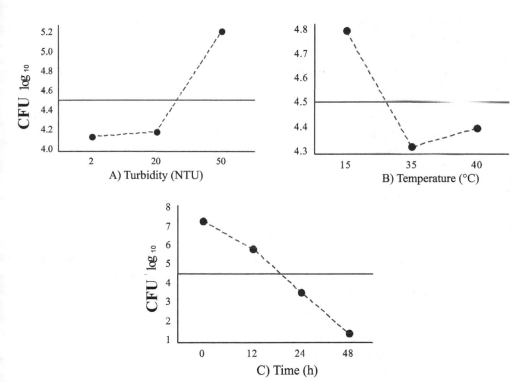

Fig. 2. Principal effects of the three abiotic factors on *Salmonella* survival. A) Different levels of the turbidity factor, B) Different levels of the temperature factor, C) Different levels of the time factor.

Lemarchand and Lebaron (2002) mentioned that in the stationary phase, *Salmonella* spp serotypes present in the environment may be constantly changing and developing resistance to a variety of environmental factors, therefore the results of this research show a positive correlation in the survival of *Salmonella* spp with respect to survival time at 15°C being this temperature as the most suitable to sustain *Salmonella* survival.

Other studies indicate that *Salmonella* sp has different survival mechanisms, such as decreasing metabolic activity and being viable but not cultivable in the environment, using genetic changes in its structure, likewise different defense mechanisms for protection that are activated in stationary phase such as the production of protective enzymes in the cell wall have been reported (Jonge *et al.*, 2003, Santo Domingo *et al.*, 2000).

In the present study it was observed an enhanced survival of *Salmonella* in high turbidity environment. These results are consistent with those reported by Mezriqui et al. (1995) whom reported the protective effect of high turbidity when UV light action in high turbidity at the lethal action of UV light from sunlight is affected by the high turbidity found in water used for the experiment.

Listeria sp Survival

Survival of *Listeria* sp was higher at 15°C (Figure 3). However, the concentrations found in temperatures of 35 and 40°C with 4.1 and 4.0 log $_{10}$, respectively were not statistically different.

The temperature of 15°C showed higher *Listeria* survival at all three levels of turbidity. The bacterial concentrations found at 50 NTU ranged from 5.2 to 5.4 $\log_{10}$, whereas at temperature of 35 to 40°C bacterial populations decreased significantly. The turbidity of 2 NTU presented a decrease in bacterial concentrations (3.5 $\log_{10}$). In general, the temperature of 15°C and 50 NTU presented a higher advantage improving *Listeria* survival.

ANOVA analysis showed that most variability between treatments was due to the effect of all the factors presenting statistical significance (p <0.05).

The results for the temperature factor for *Listeria* sp revealed significant differences. Differences were found at 15°C, which had concentrations of 5.1 $\log_{10}$, compared to 35°C and 40°C with a concentration of 4 $\log_{10}$ and 4.1 $\log_{10}$, respectively.

15   35   40

According to the main effects plot bacterial survival between different turbidities showed differences and by the analysis of the Tukey test (p<0.05) differences were observed with respect to the highest turbidity.

2   20   50

The factor time had a decreasing variability effect in the course of the experiment on *Listeria* sp survival. The Tukey test (p<0.05) revealed significant differences.

48   24   12   0

It was evident that *Listeria* sp could survive longer periods of time at 15° C than 35 and 40°C, since this bacterium develops at low temperatures with an optimum growing temperature of 4°C (Murray et al., 1997). Monfort et al. (2000) conducted a comparative survival study of *Listeria inocua* and *Salmonella* Panama in pond water for agricultural use,

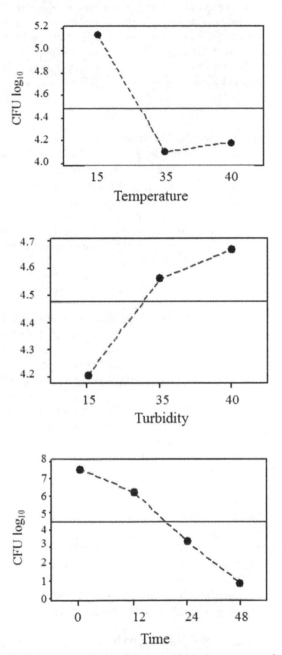

Fig. 3. Principal effects of the three abiotic factors on *Listeria* sp survival. A) Different levels of the temperature factor, B) Different levels of the turbidity factor, C) Different levels of the time factor.

where *Listeria* survival was favored at 4°C and in contrast to elevated temperatures (18°C), this data is consistent with the found in the present investigation, obtaining final concentration of *Listeria* sp at 15°C in 48 hours of 5.1 $\log_{10}$ while at high temperatures of 35 and 40°C the concentration was 4.0 $\log_{10}$. These results are similar to those reported by several studies that state that this bacterium has a higher survival at low temperatures, as optimal grow this 4°C (Szabo et al., 2003; Garrec et al., 2003).

4. Conclusions

When water comes into contact with fresh produce, the possibility of contamination from this source depends on its quality and origin. The risk of microbial contamination of water used in irrigation and processing activities should be reduced.

The results show evidence that *Salmonella* and *Escherichia coli* are present in irrigation water and can represent a potential risk for human health when they get in direct contact with edible vegetables. Therefore, control strategies must be implemented.

The survival of these bacteria was similar for the three different factors, but low temperature and high turbidity seem to represent the most suitable conditions for bacterial survival. However *Listeria* sp survival in canal water is limited.

5. References

Abushelabi Aisha A., Sofos John N., Samelis John, Kendall Patricia A. (2003). Survival and growth of salmonella in reconstituted infant cereal hydrated with water, milk or apple juice and stored at 4°C, 15°C and 25°C. Food Microbiology. 20: 17-25.

Al-Ghazali M.R, Al –Azawi Salwa K. (1990). *Listeria monocytogenes* contamination of crops grown on soil treated with sewage sludge cake, journal Applied Bacteriology, 69: 642-647.

American Public Health Association (APHA). (2001). Standard methods for the examination of water and wastewater, 18 th ed. Washintong, DC.

Banning N, Toze S, and Mee B.J. (2002). *Escherichia coli* survival in groundwater and effluent measured using a combination of propidium iodide and the green fluorescent protein, Applied Microbiology, 93: 69-76.

Beuchat, L.R. (1996). Pathogenic microorganism associated with fresh produce. *Journal Food Protection* 59: 204-216.

Bhagwat. A.A. (2003).Rapid detection of *Salmonella* from vegetable rinse-water using realtime PCR.*Journal Food Microbiology*. 607: 1-6.

Bitton, G. (1994). Wastewater Microbiology. Willey –Liss. NY

Blostein, J. (1993). An outbreak of *Salmonella javiana*Associated with Consumption of watermelon.*Journal Environmental Health.*56: 29-31

Cifuentes E., Blumental U., Ruiz P.G., Bennett S., Peasey A.(1994). Escenario epidemiológico del uso agrícola del agua residual: El valle del Mezquital, México. Salud Publica Méx. 36: 3-9

Copes J, Pellicer k, Malvestiti l, Stanchi N. (2000). Sobrevivencia en tablas de cocina de Madera y plástico inoculadas experimentalmente con *Listeria monocitogenes* , Analecta Veterinaria, 20: 47-50.

Cho Jang-Cheon,Kim Sang-Jong. (1999). Viable, but non-culturable, state of a green fluorescence protein-tagged environmental isolate of *Salmonella typhi* in groundwater and pond water, FEMS Microbiology Letters, 170: 257-264.

Diaz-Sobac, R, y Vernon-Carter, J. (1999). Inocuidad de microbiológica de frutas frescas y mínimamente procesadas. Ciencia y Tecnología de Alimentos de Galicia. 2(3): 133-136.

Fernández E.E. (2001). Algunos problemas en torno a la evaluación de agentes antimicrobianos utilizados en el tratamiento de frutas y verduras crudas. XXXII Congreso Nacional de Microbiología. 43 suplemento 1. p 54

Glesson, C. And Gray, C. (1997). The coliform index and wastewater disease. E. and FN Spon (ed.). U.K.

Guo X, Marc W., Iersel V., Chen J., Brackett R.E. and Beuchat L.R. (2002). Evidence of Association of *Salmonella* with Tomato Plants Grown Hydroponically in Inoculated Nutrient Solution..Ampplied and Environmental Microbiology. 68: 3639-3643.

LemarchandKarine, Lebaron Philippe. (2002). Influence of mutation frequency on the persistence of *Salmonella enterica* serotypes in natural warter, FEMS Microbiology Ecology, 471: 125-131.

Loaharanu P. (2001). Creciente demanda de alimentos inocuos, la tecnología de las radiaciones constituye una respuesta oportuna. Boletín FAO/OIEA 43/2.

Mezriqui N, Baleux B, And Troussellier M. (1995).A Microcosm Study Of The Survival Of Escherichia Coli And Salmonella Typhimurium In Brackish Water, Pergamon, 29: 459-465.

Monfort Patrick, Piclet Guy, and Plusquellec Alain. (2000). *Listeria Inocua* And *Salmonella Panama* In Estuarine Water And Seawaret : A Comparative Study, Pergamon, 34: 983-989.

Monge R., Chinchilla M., Reyes L. (1996). Presencia de parásitos y bacterias intestinales en hortalizas que se consumen crudas en Costa Rica. Applied and Environmental Microbiology. 369-375.

OkafoCordelia N, Umoh Veronica J, GaladimaMussa. (2003). Ocurrence of pathogens on vegetables harvested from soils irrigated with contaminated streams, The Science of the Total Enviroment, 311: 49-56.

Rice E.W, and Jhonson H. (2000). Short Comunication :Survival of *Escherichia coli* O157:H7 in Dairy Drinking Water, Environmental Protection Agency, 83: 2021-2023.

Roszak D.B., Colwell R.R. (1987). Survival strategies of bacteria in the natural environment. Microbiological Reviews. 365-379.

Siller-Cepeda JH, MA Baez-Sañudo, A Sanudo-Barajas, R Baez Sanudo. (2002). Manual de buenas pracicas agricolas. Guia para el agricultor. Centro de Investigacion en Alimentacion y Desarrollo, A.C.. SAGARPA. 1ra (ed). 62 pp.

Siller-Cepeda JH, C Chaidez-Quiroz, N Castro-del Campo. (2009). The Produce Contamination Issues in Mexico and Central America. In: The Produce Contamination Problem: causes and solutions. Sapers GM, EB Solomon, KR Mathews. Elsevier. pp 309-326.

Solomon B Ethan.,YaronSima, Matthews and Karl R. (2002).Transmission of *Escherichia coli* O157:H7 from Contaminated Manure and Irrigation Water to Lettuce Plant Tissue and Its Subsequent Internalization Applied and Environmental Microbiology, 397–400

Tyrrel S.F, and Quinton J.N. (2003). Overland flow transport of pathogens from agricultural land receiving faecal wastes, Applied Microbiology, 94: 87s-93s.

http://www.inegi.org.mx/sistemas/mexicocifras/default.aspx?ent=25

Permissions

The contributors of this book come from diverse backgrounds, making this book a truly international effort. This book will bring forth new frontiers with its revolutionizing research information and detailed analysis of the nascent developments around the world.

We would like to thank Dr Iker García-Garizábal and Dr Raphael Abrahao, for lending their expertise to make the book truly unique. They have played a crucial role in the development of this book. Without their invaluable contribution this book wouldn't have been possible. They have made vital efforts to compile up to date information on the varied aspects of this subject to make this book a valuable addition to the collection of many professionals and students.

This book was conceptualized with the vision of imparting up-to-date information and advanced data in this field. To ensure the same, a matchless editorial board was set up. Every individual on the board went through rigorous rounds of assessment to prove their worth. After which they invested a large part of their time researching and compiling the most relevant data for our readers. Conferences and sessions were held from time to time between the editorial board and the contributing authors to present the data in the most comprehensible form. The editorial team has worked tirelessly to provide valuable and valid information to help people across the globe.

Every chapter published in this book has been scrutinized by our experts. Their significance has been extensively debated. The topics covered herein carry significant findings which will fuel the growth of the discipline. They may even be implemented as practical applications or may be referred to as a beginning point for another development. Chapters in this book were first published by InTech; hereby published with permission under the Creative Commons Attribution License or equivalent.

The editorial board has been involved in producing this book since its inception. They have spent rigorous hours researching and exploring the diverse topics which have resulted in the successful publishing of this book. They have passed on their knowledge of decades through this book. To expedite this challenging task, the publisher supported the team at every step. A small team of assistant editors was also appointed to further simplify the editing procedure and attain best results for the readers.

Our editorial team has been hand-picked from every corner of the world. Their multi-ethnicity adds dynamic inputs to the discussions which result in innovative outcomes. These outcomes are then further discussed with the researchers and contributors who give their valuable feedback and opinion regarding the same. The feedback is then collaborated with the researches and they are edited in a comprehensive manner to aid the understanding of the subject.

Apart from the editorial board, the designing team has also invested a significant amount of their time in understanding the subject and creating the most relevant covers. They scrutinized every image to scout for the most suitable representation of the subject and create an appropriate cover for the book.

The publishing team has been involved in this book since its early stages. They were actively engaged in every process, be it collecting the data, connecting with the contributors or procuring relevant information. The team has been an ardent support to the editorial, designing and production team. Their endless efforts to recruit the best for this project, has resulted in the accomplishment of this book. They are a veteran in the field of academics and their pool of knowledge is as vast as their experience in printing. Their expertise and guidance has proved useful at every step. Their uncompromising quality standards have made this book an exceptional effort. Their encouragement from time to time has been an inspiration for everyone.

The publisher and the editorial board hope that this book will prove to be a valuable piece of knowledge for researchers, students, practitioners and scholars across the globe.

List of Contributors

Iker García-Garizábal
University of Zaragoza, Spain
Spanish Geological Survey, Spain

Raphael Abrahao
MIRARCO–Mining Innovation, Laurentian University, Canada

Jesús Causapé
Spanish Geological Survey, Spain

Francisco Mojarro Dávila, Carlos Francisco Bautista Capetillo and José Gumaro Ortiz Valdez
Universidad Autónoma de Zacatecas, Mexico

Ernesto Vázquez Fernández
Universidad Nacional Autónoma de México, México

Miguel Angel Segura-Castruita, Pablo Yescas-Coronado, Jorge A. Orozco-Vidal and Enrique Martínez-Rubín de Celis
Instituto Tecnológico de Torreón, Mexico

Luime Martínez-Corral
Instituto Tecnológico Superior de Lerdo, México

Mai Van Trinh and Do Thanh Dinh
Institute for Agricultural Environment, Vietnam

K.H. Baker
Life Science Program, Penn State Harrisburg, USA

S.E. Clark
Environmental Engineering Programs, Penn State Harrisburg, USA

Takehide Hama
Graduate school of Agriculture, Kyoto University, Japan

Carla Cassaniti and Daniela Romano
Department of Agriculture and Food Science, University of Catania, Italy

Timothy J. Flowers
School of Plant Biology, Stirling Highway, Crawley, Western Australia, Australia

Cristina Matos, Ana Sampaio and Isabel Bentes
University of Trás-os-Montes e Alto Douro, Portugal

Adrijana Filipović
Faculty of Agriculture and Food-Technology, University of Mostar, Bosnia and Herzegovina

P. Nila Rekha and N.K. Ambujam
Soil and water conservation engineering, Central Institute of Brackishwater Aquaculture, (ICAR), Chennai Center for water resources, Anna University, Chennai, India

Pilar Mañas, Elena Castro and Jorge de las Heras
Centro Regional de Estudios del Agua (CREA), Universidad de Castilla-La Mancha Spain

Nohelia Castro-del Campo, Célida Martínez-Rodríguez and Cristóbal Chaidez
Centro de Investigación en Alimentación y Desarrollo, A.C. Culiacán, Sinaloa, México

Printed in the USA
CPSIA information can be obtained
at www.ICGtesting.com
JSHW011429221024
72173JS00004B/731